T0074855

ALSO BY GEORGE MUSSER

Spooky Action at a Distance

The Complete Idiot's Guide to String Theory

PUTTING OURSELVES
BACK IN THE EQUATION

PUTTING OURSELVES
BACK IN THE EQUATION

Why Physicists Are Studying
Human Consciousness and AI
to Unravel the Mysteries of the Universe

GEORGE MUSSER

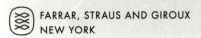
FARRAR, STRAUS AND GIROUX
NEW YORK

Farrar, Straus and Giroux
120 Broadway, New York 10271

Title-page art by Jurik Peter / Shutterstock.com.

Library of Congress Cataloging-in-Publication Data
Names: Musser, George, author.
Title: Putting ourselves back in the equation : why physicists are studying human
    consciousness and AI to unravel the mysteries of the universe / George Musser.
Description: New York : Farrar, Straus and Giroux, 2023. | Includes bibliographical
    references and index.
Identifiers: LCCN 2023018730 | ISBN 9780374238766 (hardcover)
Subjects: LCSH: Neurosciences. | Biophysics. | Quantum theory. | Cosmology.
Classification: LCC RC343 .M97 2023 | DDC 616.8—dc23/eng/20230623
LC record available at https://lccn.loc.gov/2023018730

Designed by Patrice Sheridan

Our books may be purchased in bulk for promotional, educational, or business use. Please
contact your local bookseller or the Macmillan Corporate and Premium Sales Department at
1-800-221-7945, extension 5442, or by email at MacmillanSpecialMarkets@macmillan.com.

www.fsgbooks.com
www.twitter.com/fsgbooks • www.facebook.com/fsgbooks

10 9 8 7 6 5 4 3 2 1

FOR BRET AND JON

CONTENTS

PUTTING OURSELVES
BACK IN THE EQUATION

1

THE TWIN HARD PROBLEMS

I GO TO A LOT of physics conferences, where I learn the latest about black holes, the Higgs boson, dark matter, and the deepest workings of the natural world. But about ten years ago, I started seeing an unexpected topic on the conference agendas: the mind. In the evening, when attendees gathered for drinks or dinner, I wouldn't have to wait long before physicists brought up the topic of consciousness. This was new: physics has sought for ages to get our minds *out* of the picture—to transcend our everyday experience and reveal how puny humans are compared to the vastness of the cosmos.

People sometimes told me that an intensely personal experience had awakened their interest. The young Italian physicist Giovanni Rabuffo got his PhD in 2018 in quantum gravity—the branch of theoretical physics that seeks to comprehend the nature of space and time and, with them, the origins of the universe. He was first drawn to physics as a teenager growing up in a hill town southeast of Rome. "It's so abstract, it's so precise, and it really goes down to deep things," he said. "It's reshaping philosophy. It's saying things about nature you could not discover using normal reasoning."

In 2013, when he was twenty-three and studying for his master's degree at the University of Pisa, Rabuffo heard about lucid dreaming: dreams when you know you're dreaming. Curious to experience this phenomenon himself, he found a how-to guide that advised him to start practicing meditation, and over time he learned to quiet his thoughts. Although Rabuffo never hid his side interest from his classmates, neither did he go around advertising it. "In the field of physics, sometimes I found it hard to transmit this passion, this curiosity," he said. "I found, in some people, a wall to these arguments. Not everyone: the community, in my experience, is really split." His girlfriend couldn't understand why anyone would be interested in lucid dreaming, he recalled: "She was like, 'So what?' So we broke up."

One day while he was lying in bed, Rabuffo realized he was dreaming. Excited that he might finally be having a lucid dream, he imagined thrusting his arms out from his body—whereupon he found himself *outside* his body. This didn't feel like a dream at all. "It was extremely, extremely realistic, like I was really there, awake," he told me. Rabuffo saw that his room was dark, but dimly lit by a blue light. As he began to navigate his environment, he found the source of the light: "I moved toward the mirror, and I saw in the mirror, there was not me, there was this light moving," he said. "As I was moving, the light was approaching the mirror, so I understood that that light was me." He went to the door, tried to open it, and heard the handle make a rusty sound, before feeling himself pulled back to the bed and into his body. The experience lasted maybe a minute.

Rabuffo continued with his studies and eventually relocated to Marseille, France, to finish his PhD. But he couldn't shake his fascination with what he'd experienced. He began knocking on neuroscientists' doors until a leader of the European Human Brain Project offered him a postdoctoral position, where he has found that neuroscientists value his physics expertise in coping with a flood of data. "They need that mathematics," he said. At the same time, he isn't presumptuous: "The best quality you can have as a physicist is an open mind. You don't pretend to know everything." Rabuffo hopes one day to be able to study why the brain produces sensations such as his. "It is surprising how frequent are these unconventional experiences—and how much they are disregarded," he said.

RABUFFO'S OUT-OF-BODY EXPERIENCE may be unusual, but he is hardly the only physicist in recent years to have pivoted from studying the workings of the cosmos to contemplating the intricacies of brains—both natural and artificial. Lenka Zdeborová, a Czech-born theorist who is now at the Swiss Federal Institute of Technology in Lausanne, specializes in statistical physics, which looks at the behavior of large groups of particles—billions of particles or more. It comes as no surprise that these vast hordes are complicated. What is much stranger is that their complexity begets simplicity. Through a marvelous and only hazily understood propensity for self-organization, particles spontaneously arrange themselves into collectives: crystals, gases, glasses, and other forms that are being discovered all the time.

In 2015, when Zdeborová was working on her application for a grant that required recipients to take their careers in an entirely new direction, she read about the renaissance of artificial intelligence (AI) after decades of dashed hopes. She thought back to playing chess as a teenager in the '90s and watching IBM's Deep Blue chess computer beat the human champion Garry Kasparov. Deep Blue, which relied on rules that had been painstakingly programmed into it, was the crowning achievement of traditional AI methods.[1] But in a strange way it was a letdown. The machine worked largely by brute force, and its victory, though certainly impressive, was no more surprising than a computer calculating π to a trillion places. For real insight into the nature of intelligence, most researchers in the '90s and early 2000s thought they were better off looking to the ancient Chinese strategy game of Go—now *that's* a game you couldn't mechanize so easily. It has too many possible moves. Playing it takes creativity and high-level thought, which seem quintessentially human. The idea of a computer "beating people in Go was for me like saying one day we will have power plants with nuclear fusion: that's always fifty years ahead," she said.

But as Zdeborová was preparing to give her oral presentation for the grant, Google DeepMind's AlphaGo program thrashed some of the world's top human Go players.[2] These victories marked the coming of age of an AI technology based on neural networks, which learned by doing—as a human does—rather than being programmed with rules as Deep Blue was. What gives these systems their almost humanlike (if still narrow) abilities? Raw computing power is part of the answer, but only part. "Fundamentally, we still don't know," Zdeborová said. "And so this makes it very interesting. It's the kind of puzzle that physics loves." She made such a strong case for crossing disciplines in her presentation that she convinced not only the grant agency but even herself. "Some of the criteria that you have in grant writing are purely annoying. You write something around them, and you're not really honest. But sometimes it actually helps to find the right direction for you," she said.

Like Rabuffo, she was acutely aware of physicists' reputation for barging into other fields, assuming they already know it all. "Physicists are infamously known to do that. Sometimes we are too arrogant!" she told me. So she was careful to spell out exactly what she thought physics has to offer neural network research.

Neural networks can contain billions of computing units—their "neurons," so to speak. Billions of neurons, billions of particles: they're not so different. Both neurons and particles interact with others of their kind. Particles attract or repel one another magnetically or electrically; artificial neurons fire off signals through wires, natural ones through axons. One particle might flip another upside down; one neuron might trigger another to start firing. The physical details of these interactions differ, but at an abstract level these particles and neurons are doing exactly the same thing: organizing and reorganizing en masse. "The way we train these neural networks, they really *are* systems of particles," Zdeborová said.

Systems of particles and systems of neurons are also alike in their inscrutability. It is hopeless to try to track each and every molecule in a roomful of air (hence the "statistical" in statistical physics, which describes the behavior of particles in terms of probabilities). Neural networks, too, are so enormous that you can't predict with absolute certainty what they'll do. This makes them like humans—and this isn't necessarily what their users want. Because they are too complex to program in the traditional way and must instead be taught, they can be misled. Set loose to learn from information on the internet, for instance, they absorb all its racism and sexism.[3] They have been exposed to enough human psychology to be capable of manipulating us, as we are now beginning to see with AI-powered chatbots.[4] For these reasons, achieving some deeper understanding of neural networks is essential to designing and using them wisely. "Without really new ideas, just with pure engineering power, there are things we will not overcome," Zdeborová said.

As chapter 2 will explore, the methods of physics readily carry over to neural networks. You can conduct experiments on networks as if they were gases or crystals and discover the laws governing their behavior. "It's so complicated, nobody understands," Zdeborová said. "So it's the same as if it were real nature. So it's an object for physics. It's too complicated to understand elementarily. We really need to look at [a neural network] as a product of nature and treat it as if it was a physical experimental system." She and others seek a general theory of intelligence that will apply not just to artificial brains, but to ours as well.

Lots of young physicists are following in her footsteps. "I'm very overbooked with student interest," Zdeborová said. "They just tell me, 'Oh, I fell in love with the subject. I don't know really why.'" It doesn't hurt that students see AI as a brighter career path. The physics life is hard: jobs are scarce, hours are long, progress takes decades.

There's a long history of physicists colonizing other fields for lack of opportunities in their own.[5]

One of the highest-profile physicists to change the direction of his career is Max Tegmark, a cosmologist at the Massachusetts Institute of Technology (MIT). I got to know him in 1998, when he was a postdoc analyzing measurements of the primordial universe. Later we worked together on an article in *Scientific American* arguing that our universe is only one among many—that we live in a vast, perhaps infinite, multiverse.[6]

During an afternoon coffee break a few years ago, Tegmark told me: "When I was a teenager and I realized I loved mysteries—the bigger they were, the more I really loved them—it felt to me that the two greatest mysteries of all were our universe out there and the universe in here." He tapped his forehead. Tegmark had devoted the first twenty-five years of his career to the former universe because it seemed premature to tackle the latter. Now, consciousness—specifically what philosophers call phenomenal consciousness, the nature of our subjective experience—strikes him as ripe for progress. "It doesn't feel premature anymore," he said. "It feels like cosmology's peaked in some ways."

Tegmark, who has been applying cosmological data-analysis techniques to brain imaging, thinks scientists can begin to gain traction on consciousness through integrated information theory, which contends that the brain is conscious to the extent that its parts act together in harmony. (Chapter 3 will delve into this theory.) The cerebrum's vast network of neurons works as a unified whole, fusing sights, sounds, and memories into a seamless field of experience. Its cohesiveness is not unlike the collective order that Zdeborová finds in particle systems, and Tegmark thinks the same mathematical methods that describe those systems should apply to the brain, too. "There's probably some equation: if information processing obeys *this*, there's an experience; otherwise not," he said.

Lest you think he has neglected the big picture, Tegmark thinks of intelligence as a cosmic phenomenon. Like the late theoretical physicist Freeman Dyson, he believes that in the far future, our distant descendants may be an astrophysical force on a par with natural ones.[7] So, cosmologists who make predictions about the fate of the universe need to consider the goals and abilities of intelligent beings. But Tegmark is worried this may be a moot point if humanity can't survive the many existential perils it has created for itself, such as nuclear war and superintelligent robots. (He's also concerned that we may not have seen any extraterrestrial civilizations because they all blew themselves up—not a promising precedent for us.) Tegmark notes that scientists have expertise on these threats and, to be frank, have played a role in creating them. "We therefore have a special responsibility to weigh in to counterbalance 'fake news' and 'alternative facts,'" he said. In 2014 he cofounded the Future of Life Institute to that end.

These physicists still consider themselves physicists. They don't feel that they have left the field, just that they are pursuing it by other means. Not only do they think they might be able to help neuroscientists and philosophers of mind, they also think researchers in other disciplines can help *them*. As we'll see, the latest advances in physics present scientists with a paradox: We can't understand the measurable, material universe beyond our minds without first understanding our minds. Physics seeks objective reality, but can't get away from the subjective element. As Tegmark put it: "If you look at the problems that we're still stumped on in foundational physics, pretty much all of them trace back to consciousness."

A "TICKING TIME BOMB"

Physics is the science of hard, elemental stuff, while the mind is messy and mushy—it can't be captured in an elegant equation or graphed using x and y axes. You can write the equations describing the behavior of light in a few lines and derive all the principles of optics from them; at the same time, you can read a thousand-page novel and still not feel you understand the characters, or spend your whole life in therapy and still not fully know yourself. Physicists have traditionally left subjective experience to psychologists, poets, and pastors, and those specialists, in turn, have had little use for physics.

This disciplinary divide isn't temporary or easily overcome; it is inherent in the four-hundred-year-old trade-off that opened the way for modern science. Physics was born from the split between mind and matter. The scientific revolutionaries of seventeenth-century Europe, notably Galileo Galilei and René Descartes, defined their domain as what is externally observable and quantifiable.[8] Basically, that meant motion: the paths of cannonballs, planets, pendulums, and so on. You can measure them, plot them, mathematize them. The study of motion, known as mechanics, is still the first thing that physics students engage with. When I was in college, I actively resented that. I wanted to get to relativity theory, quantum fields, the big bang—the juicy stuff. But with age I have come to appreciate the simple things. Swing a pendulum or throw a stone in a pond, and you are already demonstrating essential concepts of advanced physics: oscillation, momentum, energy. The genius (or luck) of Galileo and Descartes was that almost the entire physical universe can be analyzed in terms of movement.

These early scientists weren't uninterested in the functioning of the mind—Descartes was as much the founder of cognitive science as of physics[9]—but they recognized it would be harder to explain, and so effectively bracketed it. This split enabled the divide-and-conquer strategy that made science so successful. But the split between subjective and objective was also, in the words of the University of Toronto philosopher William Seager, a "ticking time bomb," because some things just can't be conquered by dividing them.[10] Clearly, any investigation of the brain requires scientists to integrate mind with matter. Less obviously, so do the problems in foundational physics Tegmark was alluding to, in which our understanding of matter hinges on the nature of consciousness. Scientists could kick these interdisciplinary problems down the road for only so long.

For today's physicists, nothing forces the issue more than the puzzles of quantum mechanics. Quantum theory underpins our modern description of matter. It governs everything from DNA mutations to supernova explosions and enables technologies from transistors to lasers. No exception to quantum mechanics has ever been found. But there is a troubling superficiality to its success: scratch the surface, and the theory makes no sense. Put three physicists in a room, and you will get four ideas about what it means. I know, because I've been in many such rooms. Once I was at a formal dinner with a panel of Nobel laureates and other distinguished physicists and philosophers who had gathered to debate the meaning of quantum mechanics; some had flown halfway around the world for the occasion. But they were so divided that they ended up talking mostly about international economic development instead. Consider how divisive quantum mechanics must be in order for politics to be the more polite subject!

What really riles up quantum physicists is that conscious observers seem to play an essential role in quantum theory—we seem to shape reality by looking at it. Of course we affect reality to some extent just by living and breathing, but quantum mechanics goes way beyond that. Suppose we want to measure the position of a particle. In classical (prequantum) physics, we presume that the particle is somewhere in the lab and that our instruments will tell us where. Those instruments will probably disturb the particle a bit, but with better engineering we can minimize that; there is no theoretical limit to the finesse of our experimental equipment. In quantum mechanics, the measurement process is much less intuitive. The quantities we measure take on their values only when we conduct the measurement; before we do, quantum theory tells us those quantities are undefined, blanks not yet filled in. A particle may start off not having a definite location, not sitting anywhere in particular. But when you go to look for it, lo and behold, you find it in some specific place. Had you not gone to look for it, the particle would have remained in its ambiguous state.

What is more, quantum theory says the position of the particle cannot be pinned down by a piece of equipment registering it mechanically; that merely transfers the ambiguity from the particle to the equipment. Now the measurement equipment, too, no matter how sophisticated, is in a muddle of not detecting the particle in any one place. As physicists realized in the early 1930s, there's only one thing we know of that unambiguously discerns the particle in one place or another and thereby fixes its properties: the *mind* of the observer.[11] This is strange and unsettling. And in addition to failing to explain why sentient observers should have this godlike power, the theory doesn't even spell out what an observer is. As one physicist quipped, Does an amoeba count? A human? Any human? Does the human need a PhD?[12]

Reaching for an explanation, most physicists latch on to the word "seem." Sure, it may *seem* that observers play this special role, but they don't really. Somehow we are misconstruing the theory. Maybe the ambiguity is an artifact of our own limited understanding rather than a fact of nature—the particle has a position all along, even if the theory doesn't capture it. Or maybe observers are just a stand-in for something else that is more solidly defined—the particle might need to interact with anything substantially bigger than itself, not necessarily a conscious being. As I will return to in chapter 4, the debate around these and other options has been logjammed for the better part of a century. Physicists need fresh ideas.

THE INSIDE/OUTSIDE PROBLEM

Another place where you hear a lot of discussion of observers is at conferences devoted to cosmology. If you had to vote for which branch of science is least likely to have anything to do with the mind, you'd probably pick cosmology. The universe is shaped by mindless, elemental forces, and on the scale of entire galaxies, the human brain is hardly a speck of dust. Yet ironically the universe's vast size is the very reason that cosmologists these days contemplate the brain.

The universe probably extends much farther than the outer limits of our vision, perhaps infinitely far. Cosmologists think most of it looks nothing like the part that is close enough for light to have reached us over the past 13.8 billion years. The universe's visible volume may be a violent place, filled with radiation, asteroids, and countless other hazards, but it is downright homey compared to what lies beyond view. Because of a process called cosmological or cosmic inflation, most of the space out there is filled with a peculiarly destructive form of energy, and not so much as a star or planet can form.[13] Our corner of outer space is hopelessly unrepresentative of the whole, so we have to be careful when we draw conclusions from it about the cosmos in general.

We live in one of the universe's rare comparatively habitable patches for the simple reason that there is no other place where we could live. What we see is skewed by our very existence—statisticians refer to this phenomenon as an observer selection effect or survivorship bias. It's common in everyday life, too: historic buildings seem so sturdily built that you are tempted to exclaim, "They don't make 'em like they used to," but you're only seeing the ones that lasted. To correct for this bias, cosmologists must spell out what a universe needs in order to contain minds able to perceive it. The requirements should be independent of the specifics of life as we know it because minds elsewhere may not inhabit bodies like ours, and even if they do, their bodies need not be carbon-based.

Very few scientists are claiming that our minds play, or even seem to play, a direct, physical role at the level of the cosmos. If Dyson is right, one day our descendants will rearrange stars and mine black holes for energy, but for now we're still just specks of dust. In this respect, the puzzles of cosmology and quantum mechanics are very different. In other ways, though, they are quite similar. Both hinge on a disconnect between what physics says is out there and what we see. In the quantum zone, physics says particles sometimes have no position, whereas we always see them as having one. In cosmology, physics says most regions of outer space seethe with lethal energy, yet we see it as almost sublimely empty.

And as subsequent chapters will explore, this discrepancy occurs across other branches of fundamental physics, too. Physics says that time has no directionality, yet to us it marches forward. Physics says causation is a fiction, yet cause and effect are evident every time we flip a switch and make the light come on. Physics says all is atoms and void, yet the world is much more ornately structured to our eyes than such a bare-bones picture would suggest.

Maybe our theories are wrong. But they hold up so well in other respects that some think the problem has to do with that ticking time bomb. Physics theories are written in the third person; they aim to represent the world as it is, standing outside any one observer's perspective. But our observations are necessarily all from our own, first-person perspective—filtered through our senses, our habits of thought, our bodily limitations, and the simple fact that in being part of the system we study, we are unable to stand outside it. Very often, these two perspectives don't align, and we get the above situations where theory conflicts with what we see. The influential German Enlightenment philosopher Immanuel Kant famously argued that we have no direct access to reality; we may perceive the world a certain way because it is the only way we *can* perceive it.

There's no generally accepted term for the relation between first- and third-person perspectives, or for their failure to align—which is strange, because philosophers love naming things. I'll call it the "inside/outside problem." It is one aspect of the subject known as epistemology. Many eminent twentieth-century physicists have argued that epistemology is as much the purview of physics as of philosophy. Albert Einstein developed his theories of relativity by thinking about how observers embedded within the world make measurements of it. By analogy, at a concert you may think you are clapping in time with the music, but the drummer may think your rhythm is off—and both of you may be right, because sound takes time to travel. An absolute notion of "simultaneous" presumes an unattainable God's-eye view. To make progress, we have to let it go.

Quantum theory forced an even deeper rethinking of epistemology. In 1948 Erwin Schrödinger said: "The scientist subconsciously, almost inadvertently, simplifies his problem of understanding Nature by disregarding or cutting out of the picture to be constructed, himself. . . . It leaves gaps, enormous lacunae, leads to paradoxes."[14] A few years later, Werner Heisenberg wrote: "The familiar classification of the world into subject and object, inner and outer world, body and soul, somehow no longer quite applies."[15] John Wheeler, a pioneering gravity theorist and cosmologist, put it this way in the 1970s: "The observer is as essential to the creation of the universe as the universe is to the creation of the observer."[16] Yet these sentiments were little more than just that—sentiments. Physics needs to make room for observers and perhaps for their conscious experience: Yeah, we get that. But how?

Physicists have tended to form their views on the observer question or inside/outside problem in a bubble, based on their own hunches about the mind, without bothering to ask for help from the experts in philosophy or neuroscience. That insularity is now breaking down. No one is quite sure what will happen as physicists and neuroscientists explore the uncharted territory that has always lain between them.

THE HARD PROBLEM OF CONSCIOUSNESS

When you talk to physicists about consciousness, it's only a matter of time before they mention David Chalmers, a philosopher now at New York University. In a much-cited 1994 lecture, he introduced a new phrase into the language of science: the "hard problem."[17]

By "hard," Chalmers meant impossible. Science as we now practice it, he argued, is inherently unable to explain consciousness; it can't get any traction on this aspect of the inside/outside problem. He pointed out that this is a failing not of any specific neuroscience theory, but of theorizing in general. Scientific explanation is reductionist. It treats material objects like IKEA furniture: you lay out their parts, bolt them together, and pray the box isn't missing some essential piece. Science gives you the assembly instructions: it tells you how the parts relate to one another and how they interact to create something larger than themselves. And even outside what we commonly think of as science, humans typically understand things by breaking them down into parts and relating each part to the others.

Chalmers suggested that a reductionist style of explanation doesn't work on the mind. The brain consists of parts—neurons and associated cells—and in principle you could trace all the signals passing among them (although, to borrow from George Eliot, if we could hear the billions of miniature lightning bolts zapping through our heads, we should die of that roar). But how does that quantifiable activity relate to our inner experience? Our experiences of the beguiling scent of a rose and the awfulness of fingernails on a chalkboard are not decomposable into smaller pieces. These experiences are also impossible to grasp by reference to anything else. To convey scarlet to somebody who has never seen that color, where would you even begin? Is it to crimson what the smell of an orange is to that of a grapefruit?[18] You can be as evocative as you want with your metaphors, but ultimately others will need to see for themselves.[19] The qualities of experience, known as "qualia" ("quale" in the singular), can't be grasped intellectually. They must be experienced firsthand.

The inadequacy of our explanatory framework for the mental realm goes back to the mind-matter split, when physics and science in general were defined as that which is not mental. "Perception and that which depends on it cannot be explained mechanically, that is, by means of shapes and motions," the German philosopher Gottfried Leibniz wrote in 1714. "And if we suppose that there were a machine whose structure makes it think, feel, and have perception, we could imagine it increased in size while keeping the same proportions, so that one could enter it as one does with a mill. If we were then to go around inside it, we would see only parts pushing one another, and never anything which would explain a perception."[20] For most of Leibniz's contemporaries, this posed no problem, because they still accepted the mind as a separate category. They had no expectation that physical science had to accommodate subjective experience.

But by the mid-1800s the success of science had encouraged a different view: forget mind; matter is all there is.[21] Especially in the wake of Charles Darwin, the mind began to seem like an un- necessary, even supernatural add-on. It must somehow arise from the workings of matter. In the 1870s neuroscientists began to study animals and people with neurological conditions, notably epilepsy, to see which parts of the brain were responsible for consciousness, a line of thinking that inspired Sigmund Freud to explore the depths of our interior lives.[22] But many felt that neuroscience methods just didn't—and couldn't—get at the deepest questions.

Chalmers updated these worries for modern scholars at a propi- tious moment, when consciousness research was flowering as never before. And telling physicists that something is impossible is sure to get them interested. Not only will they want to prove you wrong, but proofs of genuine impossibility are themselves profound insights into nature. Entire branches of physics are built on understanding why it is that we can't do things like build a perpetual motion ma- chine or travel faster than light. Other times, theories say that some- thing is impossible when it clearly isn't, which betrays a subtle flaw in those theories. For nearly a century, fundamental physics has been stuck on the problem of creating a unified theory. It seems impos- sible to unite our two bedrock theories, quantum mechanics and

Einstein's general theory of relativity, which describes gravity; they are too dissimilar. For instance, they involve different conceptions of time, as chapter 8 will discuss. But it *must* be possible to unify them, because we observe nature to be a unity; the conflict between our theories does not appear out in the real world. The seeming impossibility just means that some new big discovery awaits us. Most physicists think it will all make sense one day.

The same goes for consciousness. We are composed of matter, and we are conscious, and therefore it must be possible to reconcile those two facts. It may seem impossible, but that just means we're missing something, and neuroscience on its own may not be able to fill it in. Physicists have a track record of taking things that seemed utterly different, such as matter and energy or electricity and magnetism, and showing they are essentially the same. Can they do that for matter and mind? Although the hard problem of consciousness originated in neuroscience and philosophy, it is equally an issue for physics, because it means that at least one phenomenon still lies outside our present scientific framework. This almost guarantees that a new scientific revolution awaits us. And when that revolution comes, physicists may find the answers to those other inside/outside problems that hinge on consciousness. "We won't have a theory of everything without a theory of consciousness," Chalmers routinely tells physicists when he gives lectures to them.[23] Having cracked open protons and scoured the skies for things that current theories can't explain, they are humbled to learn that the biggest exception of all may lie in our skulls.

Admittedly, Chalmers plays to a physicist crowd. Theoretical physicists like to entertain crazy ideas, and are resigned to the fact that the rest of the world considers them crazy. (I had a friend in graduate school who told a girl he met at a bar that he was an astrophysicist. She immediately got up and walked away.) So along comes Chalmers, telling them that neuroscientists need crazy ideas—keep 'em coming. What's not to like about that?

It flatters physicists that consciousness should fall into their academic inbox—that explaining our inner experience might require new physics. Perhaps mind is a basic ingredient of nature, and just as particles have mass or electric charge, they also have mental properties. Or perhaps there is some new fundamental law, in addition to the laws governing how objects move, that determines whether they have inner experience or not. As the coming chapters will lay out, down that path lies panpsychism, the ancient idea that all the world is sentient, so physics-inspired theories of consciousness tend to be panpsychist. A lot of neuroscientists are unhappy with panpsychism. To them, it doesn't explain consciousness so much as deny that it has an explanation and, even if true, would not solve the puzzle of why our conscious experience has the specific structure we observe it to have.[24] Most neuroscientists I have spoken to prefer to think that consciousness is a specialized function limited to humans and a few other animal species, and that even humans might lack consciousness had evolution unfolded some other way.[25] If so, the answers lie in the biology of the brain, not in any broader revision to our understanding of nature. Physicists are certainly invited to pitch in on data-crunching, but should otherwise keep their crazy ideas to themselves.

Most physicists I've talked to see both sides of this debate. They don't think consciousness can be as straightforward as adding mental properties to their fundamental theories, and they defer to neuroscientists as the ultimate authorities, yet they are struck that the mind poses foundational questions. If the mind calls for a new style of explanation that goes beyond reductionism, physicists definitely have a professional interest in that.

THE OTHER HARD PROBLEM

In one chapter of his 1996 book *The Conscious Mind: In Search of a Fundamental Theory*, Chalmers mentioned a second hard problem. Not as widely known, it is slowly getting more attention; as if the hard problem of mind weren't enough, there is also the hard problem of matter.[26]

Whereas the hard problem of mind comes from neuroscience and philosophy of mind, the hard problem of matter is internal to physics. Physics describes how things are put together—the relations among their parts. That's why, as I mentioned above, it is mismatched to the mind: conscious experiences can't be reduced to parts and can't be described in relation to anything else. For the same reason, physics provides an incomplete picture even of material things. The laws of physics tell you what things do, but not what they are. A quantity such as mass or electric charge tells you how an object will speed up or change direction, yet the nature of the object slips through your fingers.[27]

The hard problem of matter is not new; it goes back, again, to Leibniz.[28] Physicists are mostly unaware of this history, but they recognize the problem from their struggles to comprehend quantum mechanics and the more advanced theory that is based on it, quantum field theory. Quantum particles have so little individual identity that you can swap two of them and absolutely nothing changes. Even to call them particles paints a picture of localized scraps of material that isn't really justified. Two or more of them can also bind together in such a way that they lose whatever innate properties they had and exist only in relation to one another. Everyone who works in this area has their own idea about what the fundamental objects of quantum theory really are, or even whether there are any, and experiments cannot help us decide among these options. So we are in the strange position of not knowing what we are talking about.

Like the inability of physics as currently conceived to describe the mind, this difficulty is not a small bump in the road, but the consequence of a basic trade-off that physicists make. It comes, first, from their reliance on mathematics. The number two could refer to two apples, two teardrops, two galaxies: the nature of the object is abstracted away. Abstraction gives physics tremendous power, but at a cost.

Physicists' search for a unified theory also contributes to the problem. Since the dawn of their field, they have discovered that things that seem very different are alternative arrangements of the same few types of building blocks. The distinctness of things is a product of how those building blocks behave rather than of their intrinsic nature. A table feels solid, but is mostly empty space; an apple is red, but its protons and electrons are colorless. Solidity and redness result from how those building blocks repel one another and interact with light. The building blocks themselves have hardly any properties; it is what they do, rather than what they are, that explains what we see. And every time physicists zoom in to a finer grain of building block, the building blocks lose a few more properties. Physicists have reached the point where almost nothing is left. One candidate unified theory, string theory, holds that the fundamental constituents of the universe have no fixed properties at all.[29]

Maybe that's OK. Some physicists and philosophers have concluded that, deep down, the entire category of "thing" is misguided and only relations exist. Things may emerge at higher levels without having any fundamental existence. But reducing everything to relations seems to lose the stuffness of stuff. How can there be relations if there's nothing to relate? Isn't that like a tryst without lovers?[30]

The hard problem of matter—this mystery of what physics is ultimately investigating—is intriguingly similar to the hard problem of mind. A purely relational description of matter leaves a universe that consists of nothing. Likewise, a purely relational description of the mind omits the experiential quality of experience—our sense that our mental life isn't reducible to information, that it has some extra fizz to it. Chalmers and others have speculated that the twin hard problems are linked and that the nature of matter is related to the nature of mind. The details are hazy at the moment, but the takeaway message is that physics needs to reach outside itself to answer its most fundamental questions.

For all these reasons, physicists are beginning to take up neuroscience and AI—in essence, they have no choice. What a robust theory of consciousness will look like, scientists don't yet know. But they are in the midst of building and testing concrete models that illuminate the mind and its relation to the universe. Their enthusiasm has infected me, so over the past several years I've been talking with physicists, neuroscientists, AI researchers, and philosophers, trying to learn more about this strange new territory at the intersection of their fields. In the following chapters, I'll dive into these puzzles, starting with what physicists might be able to say about the mind and building up to what a theory of the mind might tell them about physics.

2

THE NEURAL NETWORK REVOLUTION

JOHN HOPFIELD REMEMBERS THE WISPS of silicon smoke. It was early in 1985, when he was a Caltech professor and newly famous for proposing a novel type of artificial neural network, and he was visiting his coauthor and fellow physicist David Tank at Bell Labs to see what they could do with it. The two of them realized that the network could breeze through problems that ordinary computers choked on, and they were eager to give it a try. "We thought, Why not build one?" Hopfield recalled. "Why don't we build one *tonight*? I wasn't doing anything; David wasn't married yet."

Hopfield and Tank got to the lab at eight o'clock in the evening. They started with an empty circuit board, the size of a pad of paper, on which they could mount electronic parts. Then they rummaged through drawers. Operational amplifiers, or op-amps, would be the neurons of their artificial brain. Figuring they could wire together the network straight from the math, Hopfield didn't bother drawing a circuit diagram or adding parts to limit the electric current in event of a wiring mistake. "Protection resistors were not in our macho mindset," he said. Consequently, they fried some of the amplifiers: "David would have to tell you how many op-amps we threw out. I got to be very familiar with the smell of very hot silicon." In fact, they ran out of the devices and had to "borrow" some from another lab. "My analytical capabilities were somewhat dimmed by two beers at dinner," Hopfield recalled. At two in the morning, he called it a night, while Tank kept at it. The next morning, Hopfield came in to find a graph-paper plot proving the network worked.

Neural networks have become the defining technology of the twenty-first century, but few outside the field realize the crucial role that physicists have played in inventing, refining, and understanding them. Whatever the word "neural" might suggest, these networks are inspired as much by physics as by biology. And precisely because they are treated as outsiders to the field of AI, physicists have been immune to its academic politics. The idea of neural networks goes back to the 1870s, and the first working systems were built in the 1950s.[1] But for reasons that those who were around at the time still chew over—and that had as much to do with personality conflicts as with any rational argument—AI researchers and funding agencies turned their backs on them in the '60s and '70s.[2] "It was taboo," the eminent neural-network researcher Yann LeCun of Facebook (now Meta) told me. "You could not publish papers about neural networks anymore." Hopfield, Tank, and other physicists, as well as most biologists, were happily oblivious to this checkered history. "Physicists and biologists had no stigma. Hopfield made it respectable," said LeCun.

Neural networks are webs of basic computing units—the neurons, either natural or artificial—that are wired together electrically. A good picture to keep in mind is a rail or airline route network consisting of interconnected terminals. A biological neuron is a complex little computer in its own right, but in artificial networks each unit can be as simple as an on-off switch. Each unit monitors the output of multiple other units, and switches on if enough of them are on, as if taking a vote. The output of a unit is disproportionate to its input: a large input might evoke no response, and then a small change will suddenly release an output.

That balance of insensitivity and sensitivity, which mathema-

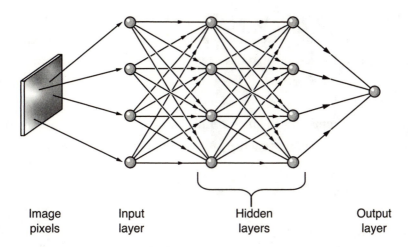

Image	Input	Hidden	Output
pixels	layer	layers	layer

2.1. FEEDFORWARD NEURAL NETWORK. A basic neural network consists of simple computing units (circles) connected by wires (lines). Engineers train it by adjusting the wiring to ensure it produces the desired output for a given input. For example, you enter image pixel data at the left, and the network outputs the probability that the image is of a cat. To make this assessment, the intermediate, or "hidden," layers typically assemble the raw data into increasingly refined geometric shapes.

ticians call nonlinearity, lets neurons serve as universal computational building blocks. String enough of them together, and you can perform whatever mathematical operation you like. Crucially, the wiring among neurons is not fixed; it can be adjusted to make the network perform the desired operation. For instance, suppose you represent the operation by drawing a curve on a sheet of graph paper. The horizontal axis is the input, and the vertical axis is the output. No matter what this curve looks like, the neural network can mimic it to a high degree of approximation. Each individual neuron fills in a piece of the curve, and a large number of them, adjusted just so and summed together, create a complete line, like snapping together Lego bricks of varying sizes to build a stairway or an archway. In this way, networks can perform operations as basic as adding two numbers or as complicated as converting image pixel data to human-readable labels such as "kitten" and "puppy."

There are several ways to build an artificial neural network. Hopfield and Tank wired together their artificial neurons using actual wires. I once built such a network using the children's electronic kit Snap Circuits.[3] On other occasions, Hopfield wrote code to simulate his networks on a standard computer, using a matrix of numbers to represent the interconnections. Researchers can also—in a fascinating new approach—culture biological neurons in a petri dish and let them link up on their own, as they would in a living brain.[4] Today, networks are usually simulated, as will be true of most of the networks I describe from here on. In each case, engineers adjust the interconnections by an automated process of trial and error, otherwise known as "training" the network. They might feed a network images labeled "kitten" or "puppy." For each picture, the network assigns a trial label, checks whether it was right, and tweaks its neuronal connections if not. Its guesses are random at first, but they get better; after perhaps ten thousand examples, it knows its pets. By the way, puppies and kittens aren't just an example I've concocted. Researchers routinely use pet pictures to test their networks—the internet has no shortage of cats.

Because neural networks are so powerful and yet so simple, they are fun to work with. That—along with the big bucks—is a big draw for many physicists. Networks collapse the distance between having an idea and showing it works. Hopfield and Tank were able to build their hardware version in a single all-nighter. Once when I was at a hole-in-the-wall restaurant in Tokyo with a group of conference attendees, one of them, eager to make a point, pulled out his laptop, tapped in a dozen lines of code, and was showing us data before we had time to order another round of sake.

That researcher—Nicholas Guttenberg, a physicist who now works at Cross Labs, a small Japanese AI research institute—encourages everybody to give it a go. Lots of apps now let you play with a simple neural network without having to write code at all. "Everyone should have experience of training a neural network, because you begin to think like an AI," Guttenberg said. "These things are idiosyncratic, and they have a kind of psychology." Knowing their quirks and their mechanisms can help to dispel *Terminator* scenarios of AIs taking over by brute force. "Part of the reason I'm not that afraid of this stuff—I don't think neural networks are going to rise up and kill us—is because I work with them," he said. The real risk, he thinks, is that people will use these systems uncritically, relying directly on their output for important decisions without checking whether it makes any sense.

Skeletal though they are, neural networks are capable of incredible complexity. Unlike most computers, which execute a program step by step, they are highly parallel systems with millions of interweaving logical paths. Each neuron acts independently: it does not have to wait its turn; it responds to its own inputs. Data flows through a network like water gushing over a waterfall, filling every void, eddying in some places and plummeting straight down in others—not by some grand design, but through droplets following whatever path their neighbors let them follow. Neural networks have an organic messiness that seems true to biological brains. They do well at the sorts of tasks our brains do well at: recognition, association, categorization. They have trouble with the tasks we have trouble with: logic, math, precision. And like us, they are not programmed, but taught.

Neural networks also bear a remarkable resemblance to the systems that physicists study. These, too, consist of simple units—particles, ions, molecules—that interact with one another and assemble into complicated structures such as crystals and glasses. We don't normally think of a snowflake crystal or a glass window as performing a computation, but, as we will see, molecular interactions are entirely equivalent to neural interconnections.

So using physics to understand these machines may tell us something about ourselves, too—not merely in the superficial sense that we are likewise made of particles, but in a deeper sense. There is a remarkable parallel between how inanimate matter organizes itself and how human perception, intelligence, and, perhaps, conscious-ness arise. Our brains, and those of our machine cousins, mirror the structure of physical reality. Mind and mindless are not such distinct categories as they may seem.

THE MACHINE IN THE GHOST

The concept behind neural networks is rooted in early efforts to reunite mind and matter not long after Galileo and Descartes split them apart. Other seventeenth- and eighteenth-century natural phi-losophers, starting with Thomas Hobbes, set out to do for the mental realm what Galileo and Descartes had done for the physical: identify simple unifying principles.[5] These thinkers zeroed in on a principle known as the association of ideas, which originated with Aristotle.[6]

Aristotle argued that the elemental mental act is the forma-tion of an association—a connection between two or more ideas, so that recalling one evokes the other. Our brains weave vast webs of associations, giving human thought its richness. Sensations pour in, eliciting some initial associations, which summon other associa-tions, ultimately leading to the actions we take and the utterances we make. Dip a madeleine into tea and memories flood your mind. Based on this old concept, Hobbes and others hoped to create a mechanistic theory of thought, both for its own sake and for the broader project of explaining the universe. Without a theory of the mind, how can we be sure that we are able to apprehend what's really out there? These natural philosophers put forward laws that governed which associations would form and when. Importantly, they argued that our brains learn these connections from experience; they do not need to be born with them. In short, they introduced the two essential features of today's neural networks: that they are cat's cradles of linked information, and that they acquire this information by taking it in rather than being preprogrammed with it.

In 1749 the English doctor and philosopher David Hartley combined this psychological theory of association of ideas with early physiological speculations by Descartes and Isaac Newton.[7] Hartley proposed that nerve tissue pulses with vibrations like ripples in a pond or sound waves in a room. As they meet one another, he thought, the vibrations reinforce or coalesce, forming physical associations that are the basis of the psychological ones. He suggested that just as individual voices blend in a chorus, simple ideas can combine to form compound ideas. Hartley conceived of memories as reverberating echoes that are spread across the brain, rather than as records that are filed into separated compartments, as others had argued.[8] In fact, modern neural networks store information in just such a distributed way.

As innovative as Hartley's theory was, he openly admitted that those vibrations were the product of speculation. Toward the end of the 1700s, scientists discovered that nerve signals are actually electrical. Studying them required the skill sets of both physics and biology. Driven, like Hobbes and Hartley, by the conviction that the brain should be comprehensible in the same basic terms as the rest of the physical world, German researchers calling themselves "organic physicists" made huge strides in the mid-1800s.[9] Their progress inspired philosophers of mind such as the Scotsman Alexander Bain and the American William James in the late 1800s.[10] Bain and James were struck by how microscope images of nerve tissue showed an interconnected network—it was as though philosophical abstractions had coalesced into physical form. In their books they drew diagrams of what we now recognize as neural networks.[11]

These networks provided a solidly grounded mechanism for the association of ideas. In Bain's model, a signal entering the network stimulates a cascade of nerve activity. If two signals enter the network at the same time, they unleash two cascades, and if the brain strengthens the linkages between the two affected groups of nerves, it will form a memory of the pairing.[12] This simple learning mechanism later got a pithy catchphrase—"Neurons that fire together, wire together"—and really does seem to occur in our brains: the chemical activity at synapses, where the output of one neuron meets the input of another, is self-reinforcing.[13] For Bain, who coined the term "stream of consciousness" to describe energy surging through the network, a conscious experience was a transition occurring within the network—the adjustment it made in response to a change in the world.[14] This hypothesis anticipated the modern theories of consciousness that I will discuss in chapter 3.

Having joined forces on these problems, physicists and psychologists parted ways again.[15] Decades later, in the 1940s, psychologists teamed with mathematicians, engineers, and physiologists to develop neural networks into a rigorous field of inquiry under the rubric of "cybernetics," a precursor of what we now call artificial intelligence.[16] Cybernetics inspired the first hardware neural networks in the '50s.[17] Although this program drew on ideas from physics, physicists weren't much involved[18]—they were making a bigger splash in molecular biology, which stands to reason, since the development of quantum physics in the 1920s and '30s made it possible to understand structures as small as viruses and DNA. Many Nobel laureates in this new field, such as Francis Crick, Maurice Wilkins, and Max Delbrück, were ex-physicists.[19] Some of these physicists turned molecular biologists later pivoted to study the brain, but the next big contribution by physics to neural-network research came from an unexpected direction: the study of gases, crystals, and other collective systems.

THE HOPFIELD NETWORK

While I was a visitor at the Institute for Advanced Study in Princeton several years ago, I often went to the cafeteria to have lunch with John Hopfield. He was eighty-five years old then and slow on his feet, but not in his thought; he remained an active researcher. Hopfield told me he went into physics because, unlike other sciences, it doesn't make students retain long lists of chemical elements, reaction mechanisms, and anatomical parts. "That was actually my strength: *not* being able to memorize," he said. "When I listened to a seminar, I was not trying to absorb the details. I was trying to see, how does this thing fit together?"

In grad school he studied how light interacts with solid materials, and in 1958 he got a job at Bell Labs in Murray Hill, New Jersey. Those were the glory days of Bell Labs—he rubbed shoulders with the inventors of the transistor, the laser, the communications satellite, and the UNIX computer operating system. He learned Go from the future Nobel laureate in physics Philip Anderson. Though hired to study semiconductors, Hopfield was encouraged to explore new fields, and at one point a colleague suggested biochemistry. Not knowing much about biochemistry, he did what many scientists in those days did when they wanted to master a new field: he taught a university course on it, as a sideline to his main job. Nothing forces you to learn like having to teach.

His first papers in biology came out of that course in 1974 and caught the attention of Francis Schmitt at MIT. Schmitt had coined the term "neuroscience"—a merger of biology and psychology—and conjured the field into existence by running a series of miniconferences.[20] To add physics to the mix, he invited Hopfield to join. Now that he was expected to give a talk, Hopfield realized he had better come up with something to talk about. "That pushed me to try to do anything they might be able to connect with," he said.

He noted that while biologists studied individual neurons and psychologists studied rats in mazes, nobody studied what came in the middle: how basic physiology resulted in outward behavior. "There were just no real connections between those people," he said. Like Hartley two centuries earlier, Hopfield thought physics could plug this gap. He inaugurated the trend that I mentioned in the previous chapter: leveraging the insights of statistical physics, which specializes in how massive numbers of particles give rise to collective effects. As billions of molecules configure and reconfigure themselves, basic properties of the world around us, such as temperature and pressure, emerge. What individual molecules do is lost in the shuffle; only their averaged behavior matters. If molecules can behave this way, why not neurons? The core functions of the brain might be collective effects, too. "When I puzzle over what is consciousness, I have a feeling, still, that it's probably emergent," he said.

Others in Schmitt's group were dubious. "If somebody has spent their life studying calcium channels, they would look at what I was doing and say, 'Your model doesn't have calcium channels. They're very important!'" Hopfield said. "To them, everything works through the details. To me, if you chase down the details, you'll never keep track of all of them; you'll never understand how it works. You're focused at the wrong level."

When Hopfield described his sources of inspiration to me, one was conspicuously absent: the neural networks that had been built in the '50s. He said he didn't think much of them. They were one-way, or "feedforward," systems, which are like computational assembly lines. Data flows in one side, it goes through various stages of processing, and the network spits out a result. Given an image, the network breaks it down and gradually works out whether it shows a puppy or a kitten. Until you give the network something, it just sits there, waiting. As he was developing his ideas in the late '70s, Hopfield doubted the brain was so straightforward. Biologists had shown it to be a "feedback" system. Its neural circuits are not one-way, but every-which-way. They loop back on themselves and are able to adapt. They need no external input to get them going. Even if you do nothing to it, a feedback system will rouse itself to activity.

To capture that dynamism in a simplified model, Hopfield fused together various concepts from cybernetics and physics—for example, the Ising model of magnetism, which is basically a description of how particles, acting like iron filings, snap into or out of alignment with one another under the influence of a magnet or even just on their own. The result was a neural network, later dubbed a Hopfield network, of an entirely different sort from the networks of the '50s.[21] The artificial neurons in it exchange signals and flip on and off in response to what other neurons are doing. One neuron might cause another to turn on, prompting others to turn on or off in a cascade effect, perhaps changing the status of the original neuron. There is no controller governing the system. The neurons develop collective patterns all on their own.

The smallest such network has just two neurons. You can set it up so that if one is on, the other goes off, and vice versa. Consider what happens if one is initially on: The other will turn off if it isn't already off. In response, the first neuron will want to turn on, but it is already on. Thus this situation—one on, one off—is a self-sustaining, stable state. When you add more neurons, the system can

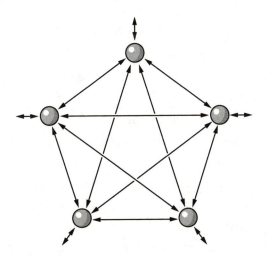

2.2. HOPFIELD NEURAL NETWORK. A Hopfield network is a type of feedback, or recurrent, neural network. The basic computing units (circles) are not arrayed in layers, but are completely interconnected. Because of this connectivity, the network has its own internal dynamics, independent of what input it may be receiving. Like other neural networks, it is trained by adjusting the wiring (lines).

have multiple stable states, or "attractors." Neurons might receive conflicting signals to turn both on and off, which will require the network as a whole to strike a compromise. It is not obvious what this compromise will be, which is why physicists find these systems fascinating.

In two important respects, the Hopfield network vindicated the early network concepts proposed by Hartley and Bain. First, its artificial neurons could be trained using the "fire together, wire together" principle, like real neurons in the brain. Second, the network stored information in a distributed form. Hopfield demonstrated both these features by training the network to act as an advanced type of computer memory.

The memory in your laptop or phone is like an immaculately organized desk where every piece of paper is properly filed into its respective folder. When you store, say, a telephone number, the computer assigns it an address, as if dropping it into one folder. To retrieve the number, the computer has to specify the address. Much of computer coding consists of moving information from one location to another and keeping track of what is where. Hopfield's neural network, on the other hand, is more like a desk covered with sprawling papers and tottering piles, whose occupant can nonetheless find anything, because the arrangement has an organic quality that captures not only a piece of information but an entire thought process.

First Hopfield input some data by manually turning neurons on or off. Then he adjusted the wiring connecting the neurons. If, say, the first and third neurons were on, he strengthened the connection between those neurons by allowing more current to flow between them. That way they reinforced each other, and the pairing of those two neurons became an attractor state of the network. If he input additional data, he could create other attractor states. The network would not lose the earlier entry; it would still be there. To retrieve information, he did not need to specify its address. Rather, he gave the network part of the data, which was enough to identify which attractor state it corresponded to, and thereby evoked all of it. With phone numbers, for example, he could give the network a few digits and it would fill in the rest. Human memory is like this. Not only can we recall information from partial clues, we can also form unexpected associations—that's almost the definition of creativity. Associative memory is also the linchpin of the neural networks that, since 2017, have revolutionized language translation such as Google Translate and chatbots such as ChatGPT.[22]

Like other systems in physics, the Hopfield network is governed by energy. This energy isn't necessarily the actual wattage consumed by the electronics over time, but a more abstract quantity represented by neuron activity. If the system meets a few simple conditions—for instance, its connections must be reciprocal, so that if one neuron affects another, it is equally affected by it—it gradually loses energy until reaching the lowest possible value, corresponding to an attractor, like water flowing down a hill and pooling at the bottom. "You started basically with a random initial state and then the network went downhill on its energy," Hopfield said. Thinking of neural activity in terms of energy was a novelty—obvious to a physicist, perhaps, but not to a psychologist, biologist, or computer scientist. "When I started in the subject, no neurobiologist ever said the word 'attractor,'" he said. "Nowadays, much of neurobiology is described in terms of attractor dynamics."

Although Hopfield intended his network to be a model of the brain, it also provided a new framework for designing computers. Computer code is usually a step-by-step logical recipe, but with Hopfield's network, you don't have to spell out those steps. The system retrieves a memory or performs some other function just by doing what comes naturally: minimizing its energy. The same tendency can also solve problems that require satisfying a bunch of constraints, from finding efficient travel routes to seating guests at a wedding reception. It is this capability that Hopfield and Tank were seeking to demonstrate during that all-nighter in the lab. Hopfield's neural network is ideally suited to such problems, which are hard because the constraints are interlocking and so must be solved for all at once.[23] You encode the constraints using the network energy. In a sudoku puzzle, for example, digits have to fit into a grid in accordance with set rules. Hopfield, always a puzzle lover, took up sudoku in 2005 and adapted his network to solve these puzzles by specifying that if two digits in a row, column, or block are the same, the energy is increased; the solution, then, corresponds to an energy of zero.[24]

In 1982, when he was first writing up his results, Hopfield discovered he wasn't the first to develop networks of this sort. Bill Little, a physicist at Stanford University, and Shun-Ichi Amari, a mathematician then at the University of Tokyo, had pioneered them a few years earlier.[25] Douglas Hofstadter, a physicist turned cognitive scientist at Indiana University, had also been arguing that thought is a collective property of masses of neurons.[26] Hopfield admitted to me that he had been sloppy about checking for prior art. Nonetheless, his 1982 paper electrified researchers in a way that previous efforts hadn't, perhaps because it reached a new audience—of statistical physicists—that wasn't conditioned to dismiss neural networks.[27] And one secret to success in research is *not* to provide all the answers. Science is a participatory sport; scientists want to work on problems they think they can help to solve. Accordingly, Hopfield left lots of open puzzles that were catnip to his colleagues. "What I contributed was really a pathway," he said. "If you were a physicist, you could see that things you knew how to do were actually useful."

THE BOLTZMANN MACHINE

"[Hopfield's] '82 paper had an enormous impact on me," said Terry Sejnowski, who is now head of the Computational Neurobiology Laboratory at the Salk Institute in La Jolla, California. Sejnowski had been a grad student at Princeton in the late '70s and studied black holes under the renowned theorist John Wheeler. He planned to write his thesis on gravitational waves—until one day a calculation gave him pause: At the rate the detectors his research relied on were improving, how long would he have to wait for experimental results? He estimated three decades. (He wasn't far off: these waves were not directly detected until 2015.)[28] "It's half a lifetime. It really pulled a rug out from under me," Sejnowski told me.

He then leapt into neuroscience, getting his foot in the door by attending an intensive summer school in 1978 at the Marine Biological Laboratory in Woods Hole, Massachusetts. He found that making electrical measurements of nerve cells came naturally to him, and gained a reputation as a good experimentalist. That made him a rarity: a theorist who didn't knock over a beaker as soon as he walked into a lab. Word got around, and a Harvard professor called to offer him a neuroscience postdoc position. At Harvard, he quickly realized how much he still had to learn, starting with humility. "It really is true that one of the things that happens when you're trained as a theoretical physicist is hubris. You are the master of the universe."

One of his early friends in his new community was Geoffrey Hinton, now at the University of Toronto. Hinton was among the few AI researchers who had stuck with neural networks through their long winter of neglect. The two men bonded over a common interest in visual perception.[29] Most of their colleagues focused on traditional markers of braininess—playing chess, proving theorems, amassing knowledge—but Sejnowski and Hinton were more impressed by the humble act of looking around. "You open your eyes; you see things," Sejnowski said. "People in AI vastly underestimated the complexity of vision."

Information pours into the brain from the eyes, yet it's somehow never enough. The world is complicated to begin with, and the retina—the detector, so to speak—provides the brain with nothing more than a flood of nerve signals. The visual cortex of the brain faces the mother of all jigsaw puzzles. It has to decide that patterns of light and dark are actually lines and textures, gradually working up to a panoply of objects laid out in 3D space. Any scene might be interpreted in multiple ways, so the visual cortex resolves ambiguities by leaning on principles of optics, geometry, and the nature of objects. "It builds up little local pieces of evidence, all over the place, into a consistent interpretation," Sejnowski said. Perception, in short, is a massive problem in satisfying constraints. So Hopfield's paper was right up his and Hinton's alley.

But as powerful as Hopfield's network was when it came to this sort of problem, it had a major flaw. The network got stuck in what researchers call a "local minimum"—a good solution, but not the absolute best. Hopfield would later notice it in his sudoku player: It could never quite finish the very hardest puzzles. It managed to fit *almost* every digit into the sudoku grid, but once it was stumped, it couldn't work backward to try again. (To be fair, that's hard for humans, too.) Sejnowski realized there was a simple fix: "I said to Geoff, 'Let's heat up the Hopfield network.'"

He didn't mean turn up the thermostat in the computer room. Rather, he was thinking of heat in a deeper sense: as random molecular motion. The molecules in a gas or other warm substance scuttle hither and thither, and what an individual molecule does is all but unpredictable. So to "heat" up the network means to add some randomness to its functioning. In 1983 Sejnowski and Hinton proposed a network whose neurons sometimes switched on or off when they weren't supposed to.[30] That sounds like a step backward. How are neurons supposed to retain memories or solve problems if they're so flaky? The point is that a little random noise can jolt the network out of a rut. It is as if you were doing a sudoku and accidentally smudged your penciled digits with the side of your hand. You'd be upset, but it would force you to think afresh.

To prove this would work, Sejnowski and Hinton reached back into the toolbox of physics, specifically the physics of heat. What is remarkable about molecular randomness is that the molecules achieve a collective order. If one speeds up, another may slow down, leaving the overall pattern of speeds—the so-called distribution— unchanged. The same goes for the neurons in Sejnowski and Hinton's network. Although they keep flipping on and off, the network as a whole settles into a steady state. Having achieved this equilibrium, the network hovers near its lowest possible energy, with occasional deviations. The nineteenth-century Austrian physicist Ludwig Boltzmann was the first to offer a mathematical description of this phenomenon;[31] the mathematician and cyberneticist John von Neumann suggested in the 1940s that Boltzmann's model would help make computers work.[32] In homage, Sejnowski and Hinton called their network the Boltzmann machine.[33]

This one simple change—adding random noise—had remarkable consequences, suggesting that Sejnowski and Hinton had latched on to an essential feature of the brain. Because the Boltzmann machine sizzles with activity, every snapshot you take of it will be slightly different. If it's performing a mathematical operation, it will return a different value each time, like a random-number generator. For many problems, that's much better than producing a single definitive answer, because it makes the machine capable of capturing diversity. The network generates a range of answers, and by tweaking the connections among the neurons, you can adjust its output to match whatever data set you're working with. For instance, if you train the network to predict the height of an adult human, it will return values from two to eight feet, with most clustering between five and five and a half feet.

When you make the network random in the same way the world is random, the network will be a simulacrum of the world. This is what AI researchers call a generative model. In recent years, the internet has been flooded with deepfakes, such as pictures of people who never existed, but look real. They are the output of a generative model (albeit of a different type from the Boltzmann machine) trained on profile pictures. Having taken in and extracted the statistical properties of human faces, the network can generate random but plausible new examples.[34] Our brains do that, too: they do not merely respond to the world, but mirror it. Not only can we look at an image and tell whether it is a kitten, we can also, if asked to imagine a kitten, picture one in our heads or, with varying degrees of artistry, draw one on paper.

In order to customize their network to a given task, in 1985 Sejnowski and Hinton worked with Dave Ackley, who was at the University of New Mexico, to develop a special algorithm—a cycle with several stages. First, you input data by manually forcing some of the neurons to turn on or off. That triggers other neurons to fire, which triggers still more neurons, and so on, until the network reaches an equilibrium. Second, you withdraw the input data and let the neurons switch on or off on their own. That triggers another cascade of activity until the network reaches a new equilibrium. Third, you compare these two equilibrium conditions. The difference will tell you how to fine-tune the network's internal connections so that it retains the data. Then you repeat the cycle for the next piece of data, and the next. In this way, the network alternates between being open to the outside world (as it receives input data) and shutting it out (as it attains a new internal balance). In short, like our brains, the network goes through a wake/sleep cycle, and for similar reasons: to help it consolidate memories. "We were totally convinced that we had figured out how the brain works," Sejnowski said.

In hindsight, the training algorithm was more significant than the Boltzmann machine itself. "The real breakthrough with the Boltzmann machine was not what I thought it would be, which was to solve constraint-satisfaction problems, but it was the learning algorithm that we discovered," Sejnowski said. The algorithm did what many AI researchers had thought impossible: train a layered or "deep" network. Inspired by it, Hinton and others developed what is now the standard training procedure for feedforward neural networks, called backpropagation.[35] It starts at the final layer and moves backward through the network, tweaking each interconnection based on how much it contributes to errors in the output. Meta's Yann LeCun later showed that this procedure, too, borrows from physics.[36]

Network layering is important because most problems have to be solved in stages. A first layer of neurons receives an input, does some initial processing, and passes the results on to a second layer, which Hinton dubbed a "hidden" layer.[37] That layer, being one step removed from the input, can operate at a more abstract level. As I mentioned earlier, a single neuron basically takes a vote: if enough of its inputs are on, it turns on. You need a hidden layer whenever the problem to be solved involves multiple rounds of voting. Here's the classic example: Determine whether two input values are both odd or both even, or one of each. You need two distinct steps. First, answer two questions: Is either odd? Are both odd? Second, compare the answers. If they differ, you must have one odd and one even number. It sounds like a lot of work for something we humans see at a glance, and that's the point. For decades, neural networks failed at this elementary task, which made Sejnowski and his colleagues' work such a breakthrough. "I still think it's the most elegant thing I've ever done," he said.

"The Boltzmann-machine paper was impressive," LeCun agreed. "I was totally hypnotized by that." He also noted that Sejnowski and Hinton still had to overcome a lingering prejudice against neural networks, and although they did speak of "networks," they avoided other trigger words: "The Boltzmann-machine paper is encoded. It doesn't mention 'neural' networks or 'neurons.' They knew the paper would be summarily rejected if they did."

"THE FIELD IS IN DESPERATE NEED"

Haim Sompolinsky, a physicist at Hebrew University in Jerusalem, told me he didn't think much of the Hopfield network when he first heard about it. I met him at a conference at the University of California, Los Angeles (UCLA), in 2019, and we chatted in the break room. "It didn't click," he said. "It was too far—we didn't understand the context." In 1984 he heard two of his colleagues excitedly discussing Hopfield's network and thought, Maybe I should pay attention.

The three of them began to explore what the network could do. "At the beginning it was really activity between physicists—having fun," he said. Neuroscientists had no interest; they doubted the network had anything to do with real brains. But Sompolinsky set out to mend this disciplinary rift: he and several physics colleagues reached out to local biologists and psychologists and began a weekly seminar. "The first phase of that encounter was horrible," he said. The biologists were dismayed that the physicists didn't know a synapse from a sulcus. They also thought the physicists were just plain weird.

For instance, Sompolinsky recalled, he and other physicists described neuron activation mathematically as either -1 for inactive or $+1$ for active. To biologists, negative numbers were the last straw. "For biologists, 0/1 was already a big simplification for them, but that's fine," Sompolinsky said. "The neuron is quiet, it's 0; active, 1. They could understand. But what does it mean, this $-1/+1$? The symmetry between $+1$ and -1—it was all very foreign to them." Worse, they thought it was unforgivable for physicists to collapse complex neuronal activity into a single matrix of numbers. "'This is just the fantasyland of physicists.' It took us three years to change the perception on both sides."

Many physicists abandoned the seminar, but a few toughed it out. "We were really focused on establishing a science," Sompolinsky said. By the early '90s, the interdisciplinary collaboration had matured enough to formalize its status, so that it could offer courses to students and apply for grants. So they founded the Interdisciplinary Center for Neural Computation. They called their nascent field computational neuroscience and devoted themselves to filling in the missing link that scholars from Hartley to Hopfield had sought: how low-level cell activity generates high-level psychology—how the mass movement of ions in your neurons translates into, say, a craving for ice cream. Physicists are especially good at bridging the gap between vastly different scales.

Today Sompolinsky thinks his young discipline is ripe for another reboot. New findings about the brain are flooding in, old models have reached their limits, and practitioners need fresh approaches. "We are now on the threshold which I believe will be a new—I hope—evolution in computational neuroscience," he said. "The field is in desperate need of that."

Many say the same about AI. During those years when Sompolinsky was helping to found computational neuroscience, AI went through another of the boom-bust cycles to which it is peculiarly prone. The excitement stirred up by Hopfield, Sejnowski, and their colleagues waned, and neural networks fell out of fashion by the mid-1990s. Even devoted champions such as LeCun stopped working on them for years. He told me there were many reasons for the downturn. Computers still weren't fast enough, software was rudimentary, and the field took a theoretical turn, focusing on computational methods that were simpler and less capable than those of neural networks, but easier to make definitive statements about. Many physicists either joined Sompolinsky in neuroscience or returned to their own field.

Seth Lloyd, a quantum-computer pioneer at MIT, told me he often hung out in Hopfield's lab while a postdoc at Caltech in the late '80s, during the neural-network slump. "I remember him saying, 'I understand that these are not working well right now, but if computers were a million times more powerful and we had a billion times the amount of data, then this would work, I swear,'" Lloyd said. He took that as a joke, but Hopfield was prescient. A factor of a million or a billion is nothing when computing power and internet

data traffic double every two years. By the mid-2010s neural networks reached human-level performance for certain narrow tasks.
"It actually happened pretty much on schedule," Lloyd said.

As Hinton, LeCun, and others finally convinced the world that
neural networks worked, the AI winter turned to spring. "I was
caught again by the neural-network storm," Marc Mézard of the
École Normale Supérieure in Paris told me over breakfast at that
UCLA conference. He is one of the physicists who had set aside neural networks in the '80s and then, in the 2010s, felt the pull again. He
figured his knowledge was hopelessly outdated and began to catch
up on reading, only to find that he wasn't as behind as he'd thought.
The theory of networks seemed almost frozen in amber. "It was
extremely striking to me that there were no new concepts. Not at
all. Everything was there in the '80s. The thing that was really new
was the experimental success."

Engineers had made systems such as AlphaGo largely by tinker-
ing, yet the lack of a deeper understanding was beginning to weigh
on them. In a much-debated 2017 lecture, Ali Rahimi, a leading AI
practitioner now at Amazon, compared his discipline to an old craft
that created lots of useful tools and methods, but was conceptually
sketchy: alchemy.[38] That was perhaps unfair to alchemy, which was
really just an early incarnation of chemistry.[39] But Rahimi's point
was that engineers needed theorists to make sense of their inven-
tions. Mézard agreed with this assessment: "The field will soon reach
a plateau—will saturate its ability to generate new devices—if new
ideas are not put in. At the moment, I find it very striking that we
have these networks that are successful at specific tasks, but we understand extremely little about how they work."

The same could be said of the brain, but the theoretical baffle-
ment about artificial neural networks is more surprising. The brain is
tightly folded inside our heads and hard for experimenters to get at,
whereas an artificial neural network is usually just a big data structure in computer memory that a user can inspect at will. "We know

absolutely everything," Mézard said, "but we don't understand what it does, how it works. This means we are not able to control how the deep network will react in a situation with an input which is very different from what it has been trained [to handle]."

So as he and other physicists flocked back to the field, they assumed a different role from the one they had before. Today, they are more explainers than inventors—they are the ones working to provide the deeper understanding of which Rahimi spoke. Lenka Zdeborová, who had been Mezárd's grad student (and had only just been born when Hopfield published his famous paper), described the shift: "Forty years ago, they were purely: 'Oh, there's a system. What can we ask of it?' But they didn't have any guide: 'Oh, there is this huge empirical success that we need to explain.' It was more exploratory. Today the motivation is much clearer."

Historically this is a common pattern for physics: tinkerers create, scientists explicate—in that order. Zdeborová gave the example of bridges. Engineers used to build them largely by the seat of their pants. "Since Mesopotamia we have known how to build bridges— but empirically," she said. "They mostly worked. Sometimes they fell." A theory of structural engineering came much later. "Today when we build a bridge we imagine there is a big statics computer simulation behind it," she said. With neural networks, too, technology raced ahead of theory. These networks and the computers needed to run them now exist—but will we be able to explain how they work?

THE GENERALIZATION PARADOX

The fondest wish of any good teacher is that students will *get* the material. They won't just regurgitate what they have been told. They will take it in, process it, and reach their own distinctive understanding. They may even teach the teacher a thing or two.

What's amazing about artificial neural networks is that they learn in this same way—and no one knows why.

Suppose you train a feedforward network to classify pet images as "kitten" or "puppy" by giving it ten thousand labeled examples. Remarkably, not only will the network properly classify those ten thousand images, but you can also give it a new one and it will classify that one correctly, too. A network might even figure out the categories of "kitten" and "puppy" on its own, without relying on labeled examples, by looking for clusters of similar images. When researchers analyze how the network is able to generalize, they find that it does not blindly memorize the data; instead, it develops a general system for representing images. It takes various arithmetic combinations of pixel values, makes combinations of those combinations, and continues until it integrates information from across the entire image. Typically this is a series of geometric operations. The first layer of neurons looks for changes in brightness. The next layer interprets those as edges; the next, as lines; and so on, up to shapes, objects, and—for images specifically of pets—ears and whiskers.[40]

If, instead of giving the network images, you feed in text, it will come up with an organizing system for written language by breaking the text into words, phrases, and sentences.[41] The network develops a hierarchy of structure that provides a framework for understanding any image or text, not just those the system is trained on. So the system doesn't only learn. It learns how to learn. And its plasticity is impressive. Yann LeCun once told me that if space were four- rather than three-dimensional, a neural network would handle it just fine.

In practice, networks do not always succeed in ascertaining this structure, and those that do may require reams of training data, so engineers give the learning process a helpful push by building basic geometric or grammatical properties into the network. LeCun made his name by coinventing a network architecture that is innately hierarchical and thus primed to handle images without having to go through all the trouble of rediscovering the rules of geometry.[42]

Language-processing systems such as ChatGPT are so powerful because they have special layers to capture contextual cues.[43]

Ten thousand sample pictures sounds like a lot, but it isn't, relative to the size of the system. A serious neural network may have millions, billions, or, nowadays, trillions of interconnections that can be individually adjusted. It has the power to skip the geometric niceties and just rote-memorize the data. But if that were all it did, it would choke on any data it hadn't already encountered. The network would "overfit" the data, meaning that it would capture the images down to every last pixel, but fail to extract the general principles: what differentiates a kitten from a puppy. It wouldn't see the forest for the trees.

The term "overfit" comes from thinking of the network's task as fitting a trend line to data. Imagine creating a graph of outdoor temperature. You check the thermometer once an hour and put a dot on a sheet of graph paper. The temperature typically climbs in the morning, peaks in the early afternoon, and drops again, so you draw an inverted U through those dots as best you can. This curve is a kind of algorithm. For a given input (the time), it has an output (the temperature). Using the curve, you can estimate the temperature in between the hourly readings, as well as make predictions about the future. An image-classification neural network does something very similar, although the input is not a single number representing the time, but a string of numbers representing the pixels in the image. The output might be the probability of a kitten or a puppy.

The real world is complicated, however. The temperature seldom follows an inverted U perfectly. It is subject to little random wiggles—the sun passes behind clouds, the wind gusts from the north, or what have you. Overfitting means that you absolutely insist on threading the curve through every data point, creating what looks more like a snake than a U. When you do that, the curve no longer tracks the underlying meteorological trend, and you can't trust its predictions for intermediate or future times—the algorithm is not of any generalizable use. A neural network, likewise, overfits if it treats every tiny feature on the image as significant. In essence, it knows too much to be useful. Fitting data, then, is not altogether different from fitting clothes. The goal is to highlight the general contours of the body, not call attention to every bulge and hollow.

Recognizing this risk, scientists in the '80s deliberately introduced some slop into their networks—a so-called regularizer—so that they would not be overly punctilious about fitting the data and would seek an appropriately loose fit.[44] But there's a trade-off: as the network disregards small deviations from the general trend, it may miss wiggles that actually are meaningful. Part of the art of designing neural networks is striking a balance between precision and generality. Our brains face the same trade-off, which is arguably the central challenge of perception and of intelligence. In a noisy and complicated world, where essential patterns are hidden beneath a crust of

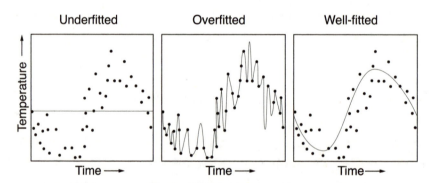

2.3. CURVE FITTING. The central challenge of intelligence—both artificial and human—is to detect essential patterns without getting bogged down in meaningless detail. For instance, thermometer readings on your front porch will follow a roughly sinusoidal diurnal cycle, with random variations due to clouds, winds, and so forth. Just taking the average temperature tells you something, but not much (underfitted). At the other extreme, reproducing each and every variation puts too much stock in irrelevant random details and can cause huge errors when making projections (overfitted). A machine system ideally strikes a balance and discerns the underlying trend (well-fitted). The same principle applies to recognizing images, processing text, and other classic machine-learning tasks.

randomness and misdirection, it's hard to tell what's important and what's ignorable. Like trees, which evolved to handle fire—they have adapted to survive all but the most intense forest fires and to thrive afterward—our brains evolved to handle chaos.

A network will make mistakes, and that's good. It *should* make mistakes on borderline cases. If it classifies a Pomeranian as a cat, you know it must be looking for general rules about what differentiates cats from dogs—their fluffiness, for example—rather than rote-memorizing all the examples it is given. Error is thus not a failure of learning, but a sign of success. A regularizer gives the network license to be wrong.

Many researchers in the '80s doubted networks would work even with a regularizer. Sejnowski told me about NETtalk, a network he and a colleague created back then to convert English text to speech. It was tiny by today's standards, with three hundred neurons and nineteen thousand interconnections, yet still seemed too big.[45] "The mathematicians laughed at us," he recalled. "They'd say, 'You're so overparameterized. There's no hope in hell. It's just going to memorize things.'" Yet that didn't happen. The machine absorbed the idiosyncrasies of spoken English and pronounced text it had never seen before. "It was able to generalize just fine," he said.

As the years passed and networks grew in size, they should have become even more prone to overfitting, yet they didn't. By the 2010s, researchers realized that networks didn't even need a regularizer; they were adept learners by their very nature.[46] Somehow they evaded the precision-generality trade-off. They could fit their initial batch of data perfectly without losing the ability to handle fresh instances—a rare case in life of being able to have it all.

These gigantic neural networks have stood statistics theory on its head. Normally, statisticians boil a lot of observations down to a few key numbers, such as the average and standard deviation. But neural networks—and big-data systems in general—take relatively few observations and extract a lot of numbers: from ten thousand images, they forge a million neuronal connections, for example. In mathematical terms, that roughly means you have ten thousand equations to solve for a million unknown parameters. Your high-school algebra teacher would draw a big red X through your homework if you tried to do that—you're supposed to have as many equations as unknowns. This suggests that the system has too little information. So how can it possibly work?

NETWORKS UNDERGO CHANGES OF STATE

This is just the sort of problem that physicists have struggled with for a century and a half. These are people who eat large numbers for breakfast. From a few key quantities such as temperature and pressure, they draw conclusions about the hundred million billion billion molecules buzzing around in a roomful of air. They may not be able to say much about individual molecules, but who needs to? The behavior of the whole system is what matters.

Central to physicists' thinking is that molecules assemble themselves into different states, such as solid, liquid, and gas, and can abruptly change from one state to another. This phenomenon is independent of the makeup of these molecules. Water, magnets, and diverse other systems all undergo changes of state in basically the same way. Physicists have even extended the principle to bird flocks and traffic jams—and now to neural networks.

Networks organize themselves collectively in many of the same ways that molecules do, meaning that they can assume different states and undergo changes of state. When water melts or boils, the connections between molecules loosen; likewise, under the right conditions, the connections between neurons can loosen, too. The networks do not literally disintegrate and flow away like a liquid, but their functioning changes in analogous ways. "It's like 273 kelvins, where you have the transition from ice to water," Marc Mézard told me.

For instance, changing the size of a network can result in something like a change of state. The specific comparison is to a granular material, like sand in an hourglass or dry cereal in a supermarket dispenser. "You've got millions of grains of sand, and there's basically an analogy with the tons and tons of parameters inside of a neural network," said Kyle Cranmer, a particle physicist and machine-learning researcher and professor who was at NYU when we spoke but is now at the University of Wisconsin–Madison. When grains are loosely packed, they can flow freely, but when they are jammed together, they get stuck and you have to bang on the side of the dispenser to release them. The freely moving condition is mathematically equivalent to a neural network that is large enough to retain all the data you give it. The jammed condition is like an undersized network that chokes on all that data.[47]

In the oversized network that is analogous to freely flowing grains, you have the situation that your algebra teacher would complain about: having fewer equations than unknowns. The equations, in this case, are the training data; you solve them to obtain the parameter values. With too few equations, there is no unique solution—but that's a good thing. The question is no longer, What is the solution? Solutions abound. The question becomes, Which solution do you pick? For those of us raised on standard statistics theory, this takes a mental adjustment. You have to stop worrying about getting *the* answer—a single optimal configuration of the network—and accept that there is a multitude of answers. Given this embarrassment of riches, the challenge is to choose wisely. Although all the solutions will describe the data you were given, only a select few will generalize to never-before-seen cases. Sejnowski puts it this way: "The degeneracy of solutions changes the nature of the problem from finding a needle in a haystack to finding a needle in a haystack of needles."[48]

It so happens that the backpropagation training algorithm that Hinton and others developed in 1986 chooses well. For reasons that are still not fully understood, it naturally homes in on just those solutions that extract the most salient lessons from the data. "We just lucked out that we happened to hit the right algorithms at that period in the '80s," Sejnowski said.

The basic principle, then, is that an overabundance of neurons may seem mathematically gratuitous, but helps a neural network to learn a task. That principle applies to the brain, too.[49] Surya Ganguli at Stanford University, another physicist turned AI and brain researcher, told me that the brain often seems to have too many neurons for its own good. A neural circuit containing a million neurons might perform a task requiring only a thousand. But the excess is essential. "You may need more neurons to learn than to actually accomplish the task after learning is done," he said. "That's happening in machine learning as well. Larger networks learn better than smaller networks even though there exists a small network that can do the task."

WHEN INFINITY IS TOO MUCH

A common tactic that physicists use when confronted with big systems is to throw caution to the wind and think of them as infinitely big. Infinity may boggle the brain, but it has a mathematical simplicity to it. If approximating very large numbers as infinity works for the theory of gases and other substances, Sompolinsky and others suggested in the '80s, maybe it works for understanding neural networks, too.[50] If networks have, in effect, infinitely many neurons, they might not be as inscrutable as they seem at first. This early research gathered dust, but was taken up three decades later by a new generation of physicists and mathematicians in their twenties and thirties. "When you take the limit of infinite width, it turns out that the problem is very simple to describe," said Yasaman Bahri. Bahri got her PhD in theoretical condensed matter physics in 2017 and went to work at the famed Googleplex in Silicon Valley, which is where I met her.

For a young scientist, the theory of neural networks is an ideal subject. The puzzles are fresh, low-hanging fruit is plentiful, and the scientific method moves at unaccustomed speed. Bahri's coauthor Jeffrey Pennington, whose doctorate is in theoretical particle physics and who works out of Google's less fancy but still impressive New York offices, remarked: "Unlike in physics—where you could have a nice idea and somebody could have done it three hundred years ago, and it takes fifteen years to come up with the idea—you can have a basic understanding of the math and statistics and get quantitative intuition, and it's easy to come up with an idea that nobody's [explored], and just do it."

Bahri, Pennington, and their coauthors imagined a system in which the input data fans out to an infinite number of neurons.[51] The signals trickle through to a second infinite layer of neurons, and a third, and so forth, and eventually are distilled into an output. The researchers initially give the network a random pattern of interconnections, so the network acts as a random number generator, except that it doesn't generate just single random numbers, but random curves.[52] If you were using an infinite neural network for the temperature example I gave earlier, it would draw a random curve across the graph paper. If you reset the interconnections, it would draw a different random curve.

This is where the special mathematics of infinity comes in. "The thing infinite width does is it makes that random behavior take on a simple form," said another coauthor, Jascha Sohl-Dickstein (PhD in biophysics), also at the Googleplex. If the network generates a bunch of these curves, a couple of patterns emerge. For a given input, the outputs tend to cluster around an average, deviating with a probability described by a bell curve. Furthermore, the system is unlikely to draw a weird, squiggly curve through the data, but will prefer a gentler arc. If you apply the network to puppies and kittens, it will choose a fairly straightforward classification rule that puts any two similar images in the same category.

To train such a network, you nudge its connections until the resulting curve passes through the available data points. In between those points, the curve takes a random path, but because of the simplicity that infinity induces, it doesn't go berserk, so the system fits without overfitting. In the end, it will come up with a sensible reconstruction that, for the example of diurnal temperature variation, is probably not far from an inverted U. This training process is easily described mathematically, and you can predict exactly what the network will do. "We are able to write down an expression for the result," Bahri said. Despite the theory's idealizations, it works well in practice, accurately predicting many aspects of neural networks that Google and other tech companies have developed.

The hope is that this infinite-network theory will offer practical tips for network design. Engineers have particular difficulty getting multilayered networks to work well; even seemingly insignificant changes to their size and architecture can render them unable to learn. In 2016 Ganguli, Sohl-Dickstein, and their coauthors highlighted one reason: the way that an input signal propagates through a feedforward network to reach the output.[53] If each layer amplifies signals even slightly, the signals will steadily intensify and surge through the system in a chaotic avalanche. In concrete terms, that means the network will take small differences and blow them out of proportion. Show it two kittens, and it will accentuate their distinguishing features until it can't learn to recognize them as the same species. If the opposite happens—if each layer weakens the signals passing through it—the network will go to the other extreme. It will collapse distinctions and be unable to detect any difference between a puppy and a kitten.

To make a system learn as well as it can learn, engineers need to keep it on the knife edge between chaotic and comatose. A slight deviation either way will turn it from an avid learner to clueless.

"Initializing at the edge of chaos can help you find better networks," Ganguli told me. Physicists call this balance a critical point, and they observe it in certain types of changes of state. Many systems in nature settle to a critical point spontaneously—including, maybe, our brains.[54] Neurons in our brains fire in groups, or "avalanches," of every size. This lack of a preferred scale is characteristic of a critical point, and it suggests a system that is pulsing on every level, ready for anything.

As powerful as the infinite-network theory is, it has limitations. Compared to the infinite number of neurons in each layer, the number of layers is paltry—in relative terms, it is as if the network has no layers at all. That is bad because layering is essential to creating abstractions. "That's not a deep network at all; that's not deep learning," said Daniel Roberts, a theoretical physicist at MIT as well as at the business-software company Salesforce. "In some sense, these infinite-width networks are behaving like they don't have any hidden layers." At a more detailed level, the randomization that occurs in each layer of an infinite-width network bleaches out information about the hierarchy of structures in an image. Such a network can't create a concept of geometric shapes, let alone tails and whiskers. If you show it two kittens, it may learn to classify them as the same animal, but only by a pixel-by-pixel comparison of the images. "The thing that we expect neural networks to do, and certainly what humans do, is build up complex representations. We think that learning representations is important. That doesn't happen at infinite width," Roberts said.

To fix up the theory, Roberts, the theoretical physicist Sho Yaida at Meta, and the mathematician Boris Hanin at Princeton University reached back into the physics toolbox and pulled out what is known as perturbation theory. It is a strategy of incrementalism. You start with a simplified case that you know how to solve, then add complications one by one. For instance, Earth's orbit is governed mostly by the sun's gravity, but Jupiter and the other planets also tug on it. So physicists first calculate the sun-centered orbit and then nudge it to account for Jupiter's influence. That creates additional long-term oscillations, including those that produce ice ages. Perturbation theory is also widely used in particle physics, where the complications in question are quantum effects or interparticle forces.

This framework translates directly to neural networks. "I'm stoked about this," Roberts said. "It's not just an analogy. It's literally the same formalism." He and his coauthors start with an infinite-width network and then adapt the solution for large but finite size.[55] When each layer is big but not strictly infinite, the number of layers begins to matter again. Each layer is no longer a randomized tabula rasa. A neuron in one layer feeds its signals to multiple neurons in the next layer and thereby causes those neurons to track one another. As signals trickle through the layers, ever more neurons are linked together, and the network integrates a wider range of information. The network becomes capable not merely of memorizing the data, but of mirroring its structure, be it geometry or grammar. This new theory provides additional practical advice for engineers, such as the optimal width for a given depth, and confirms that a neural network is not beyond rational understanding. "It's not a black box anymore," said Roberts.

QUANTUM NEURAL NETWORKS

Back in the early '80s, around the time Hopfield created his network, the Nobel-laureate physicist Richard Feynman and others hatched the idea for another transformative technology: quantum computers.[56] Their primary goal was to create a computer that would allow them to calculate the motion and dynamics of quantum particles, especially when they are subject to quintessentially quantum effects such as entanglement (which I will delve into in chapter 4). Ordinary computers, not to mention humans, struggle with these calculations. What you need is a computer that is itself based on quantum physics. Forty years later, quantum computers are mostly still physics experiments staffed by teams of PhDs, but on certain types of problems they already make the fastest classical computer look like an abacus.

Apart from being another unorthodox and exceptionally powerful type of computer, quantum computers at first glance don't seem to have anything to do with neural networks. Although Hopfield modeled his network on a grid of particles that interact by way of magnetic forces, and particles are quantum objects, he ignored that complication. "When I was first working on things, no one—not even Feynman—understood how to do quantum computing, and I always thought of things in classical terms," Hopfield said.

Elizabeth Behrman of Wichita State University told me the connection between these technologies occurred to her almost by accident. She stumbled upon Hopfield's work while a chemical-physics grad student in the early '80s. "I was reading some of Hopfield's stuff on a completely different topic . . . and just started reading his papers," she said. Years later, in 1990, she was at an orientation for new faculty when she met James Steck. Hearing that he worked on neural networks, she thought back to Hopfield's papers and was struck that many systems in physics are basically neural networks in all but name. Behrman recalled, "I was fascinated by the idea of the power of interconnectivity, and I mentioned that, of course, a solid-state system automatically has interconnectivity, since everything interacts with everything else. You don't need to wire it." And in the physical universe, that connectivity is ultimately quantum in nature.

The neurons in Hopfield's network are either on or off. But Behrman proposed "upgrading" them to quantum neurons, which could take a variety of physical forms; she and Steck focused on molecules that had two stable states, corresponding to "on" and "off." Such a neuron can behave like Schrödinger's cat: it can be both on *and* off. In fact, it can be half on and half off, mostly on, mostly off, or anywhere in between—endless possible combinations called superpositions. To be *and* not to be: that sounds like a logical contradiction. But it really just means that the state of the neuron does not neatly fall into our categories of "on" and "off," which are like trying to describe color in terms of black and white. Quantum neurons have an iridescence that we are ordinarily blind to. A network of them is in a vast superposition in which every neuron is on, every neuron is off, and some are on and some off—all at once. In the late 1990s Behrman, Steck, and their coauthors showed that this opens up vast new possibilities for information processing.[57]

Her colleagues didn't know what she was on about. "I had a heck of a time getting published," she said, "because the neural-network journals would say, 'What is this quantum mechanics?' and the physics journals would say, 'What is this neural-network garbage?'" Another quantum-network pioneer, Hidetoshi Nishimori at the Tokyo Institute of Technology, also told me his colleagues were underwhelmed. "People interested in quantum physics were interested, but it didn't go beyond that," he recalled.

Nishimori, whose field is statistical mechanics, and his colleague Yoshihiko Nonomura showed in the mid-1990s that the quantum setup eludes a difficulty I mentioned earlier—that neural networks can get stuck. They naturally settle into a low-energy configuration, but not necessarily the lowest-energy configuration, and can need a little random noise to shake them loose. That was the rationale behind the Boltzmann machine. But a quantum system has no need for added noise. It has a natural tendency to explore the possibilities open to it rather than settle down prematurely. If you put the network into some specific configuration, it will quickly enter a superposition that includes other configurations, and this superposition will rapidly grow to encompass still more possibilities. If it encounters a configuration with a lower energy, the system will nestle, or "tunnel," into that configuration.[58]

In a follow-up study in 1998, Nishimori and another colleague,

Tadashi Kadowaki, presented a procedure for using a quantum Hop-field network to solve sudoku-like problems.[59] You control the quantum exploratory behavior using an external magnetic field. You start by turning on the field full blast to initialize the neurons to an equal superposition of on and off—a blank slate. Then you start allowing the neurons to interact by their own mutual magnetic forces. Some turn on if their neighbors are on; others switch off. You don't allow just any two neurons to interact, but choose pairs or larger groupings depending on the problem at hand. For instance, if the network were solving a sudoku puzzle, you would link up neurons in rows, columns, and blocks. When one neuron in a row encodes a 1, other neurons in that row switch to a different digit. In so doing, they drive neurons elsewhere in the grid to change their values.

The external magnetic field facilitates the neuronal switching. That's good at first, but becomes a liability because sometimes the network will overwrite the right answer. So you slowly dial down the field and eventually turn it off altogether. "We could drive the state of the system from a simple state by adding quantum fluctuations and reducing the amplitude of the fluctuations to zero very gradually," Nishimori said. That lets the system reach the right answer and locks it in. This procedure is known as annealing, by analogy to a metalworking technique that relieves the internal stresses in a bar of steel by heating it and then cooling it slowly.

Facing a lack of enthusiasm among his colleagues, let alone the wider world, Nishimori turned to other topics. Only a decade later did he learn that experimenters had begun to build actual hardware based on this idea. In 2011 a Canadian company, D-Wave Systems, made its quantum information processors available for sale.[60] These aren't yet true computers (a term that connotes a fully general problem solver) and have other limitations, but are nonetheless powerful. Lockheed Martin bought one to test the computer code in its fighter jets.[61] I've heard from the company executives Vern Brownell and Mark Johnson about how newer models have streamlined supermarket logistics, unclogged traffic jams, and screened for cancer drugs. "I was very naïve in a sense," Nishimori told me. "We didn't look outside the ivory tower." His story reminded me of the documentary *Searching for Sugar Man*, about a rock musician from Detroit who was almost completely ignored in his own community, not realizing until decades later that he was a sensation in South Africa.

QUANTUM INTELLIGENCE

With quantum processors, we are witnessing the emergence of a new type of AI. These machines are unaccountably good at not only solving sudoku-like problems but also the core AI skill of learning from examples.[62] In an early demonstration in 2009, a team at Google trained a prototype D-Wave information processor to recognize cars in images.[63] A team of physicists later repurposed the same algorithm to look for the Higgs boson in particle collider data.[64]

But just because something is quantum doesn't automatically make it more powerful. The quantum annealer and the older technique of adding noise to a classical neural network are like taking the freeway and taking the surface streets. Which is faster depends on the circumstances. Consider sudoku. Sometimes you get stuck because, although you're not too far off the right answer, the only way to get there is to change some of your entries in a way that seems like a mistake. Other times, the changes seem fairly obvious, but you need to make a lot of them. The quantum annealer does well in the first situation, while noise does better in the second.

Fans of quantum computing argue that practical problems tend to fall into the first category.[65] But that claim has never been fully tested—and even if it proves to be true, why would it be so? "That's the question we keep asking ourselves," Nishimori said. "We don't know the answer yet." It may be a deep fact about the world, and a reason our own intelligence evolved to be what it is.

The advantages of quantum systems are no clearer for physicists pursuing an alternative approach to quantum machine learning. Rather than build a quantum neural network in physical form, which is what a D-Wave machine is, they simulate such a network in software on a general-purpose quantum computer. The computer translates the workings of a network into a sequence of operations that it performs on a bank of quantum bits, or "qubits." The qubits could be made from molecules, ions, subatomic particles, electrical loops, or other building blocks, but it doesn't matter; all can be programmed in basically the same way. A software-based network is more abstract, but has advantages. It identifies each neuron not with a specific particle or other hardware unit, but with a specific state of the system in its entirety. The result is a huge increase in capacity. A computer with two qubits has four states: both off, both on, one on and the other off, and vice versa. An ordinary computer could be in just one of those states at a time, but a quantum computer can be in a superposition of all four. Because those states coexist, they can serve as independent neurons. Thus you have managed to pack four neurons onto two qubits.

With every particle you add, you double the number of neurons. Three dozen qubits can give you as many neurons as there are in the human brain. This is a dramatic demonstration of the power of collective behavior in physics, of how the whole can be more than the sum of its parts. Every time you act on this system, you act on all those neurons at once. So it's no wonder that people get excited about quantum computers. But there's a catch: you need to convert the input data into a quantum superposition and, once the processing is done, translate the superposition back to a human-readable output. These translation steps can offset or even negate the machine's innate power. If quantum machine learning is faster than classical machine learning, it is not uniformly faster—it depends on the algorithm and the data to which it is applied. "It's very, very subtle where actually quantum computing is better," said Maria Schuld at the University of KwaZulu-Natal in South Africa.

In 2017 Schuld became the first person in the world to get a PhD in quantum machine learning. When I first met her, she was doing a postdoc, working at a Toronto-based startup company, consulting for Microsoft, writing a book, and creating an app to track political polarization—while still finding time to go surfing. I don't know how she does it all without being in a quantum superposition herself.

She complained that the builders of quantum computers haven't been very concerned with whether their systems are really any faster than standard techniques. They have treated machine learning as just a nail for their quantum hammer. They built their contraptions and then went looking for something, anything, to do with them. "We need to find the 'killer app'—it's called in the community—that justifies having these devices," she said. "I feel that they're almost desperately searching for it." For their part, machine-learning researchers have been content to leave physicists to their playthings. "There's very few people, actually, who are interested from [the machine-learning] community. Almost everyone who is working in the field is a quantum-computing scientist who hopes that putting machine learning on his CV gets him a better position," Schuld told me.

If quantum machine learning is to graduate from plaything to practical tool, physicists will need to explain how their systems complement existing techniques. Quantum computers would make a real splash if they made entirely new algorithms possible, showing that our existing computers and our brains do not exhaust the possibilities of intelligence. Schuld has sought to do just that. "I started to work the other way around and think: If we now have this quantum computer already—these small-scale ones—what machine-learning model actually can it generally implement?" she said. "Maybe it is a model that has not been invented yet." She and her colleagues have found ways to classify data using quantum operations that have no direct counterpart in classical machine learning.[66]

Where quantum computers are unequivocally useful is at the

task that Feynman originally envisioned for them: processing quantum data. Instead of images or text, you feed in data from physics or chemistry experiments. Then there is no need to translate, because the data is already in the machine's mother tongue. Researchers have come up with algorithms to identify quantum states, classify phases of matter, and extract key quantum quantities.[67]

A major function of intelligence is to find simplicity in complexity—not to take the world as given, but to probe beneath its appearances. That's why engineers go to such lengths to avoid overfitting. But there is no single thing called intelligence; it depends on the domain, in machines as in people. An ordinary computer, confronted with a strongly quantum reality, struggles to see any simplicity in it. To the contrary, it finds the quantum world magical and capricious. A quantum computer, on the other hand, has the right kind of intelligence for that world.

"Quantum systems very famously generate these weird and counterintuitive patterns that it's hard to bend our minds around," MIT's Seth Lloyd told me. "So the application to deep learning is: well, if they can generate these patterns that are counterintuitive and hard to bend our mind around and that we don't think can be generated by a classical computer, maybe they can recognize patterns like this as well." All in all, quantum AI systems should be thought of as different rather than superior intelligences. They may or may not help with classic AI tasks such as recognizing images and processing language, but they will one day guide us through their mysterious native realm.

BLACK HOLE COMPUTERS AND BEYOND

Physicists haven't stopped mining their theories for new approaches to neural networks. I learned about a particularly dramatic proposal from Gia Dvali, a theorist who studies cosmic dark energy and higher-dimensional spacetime. I met him at NYU, one of several institutions where he works. Whereas Hopfield modeled his network on magnetism, Dvali has suggested basing networks on the force of gravity.[68] Gravity is so weak that it scarcely operates in the brain or in a computer, but Dvali simulates what would happen if it were stronger. "Let's design a neural network as close to being as gravity-like as possible," he said.

Such a network, like Hopfield's, is governed by energy. To turn a neuron on, you have to give it some energy. But whereas magnetism can either attract or repel, gravity only attracts. So it has a self-reinforcing quality, which in a neural network means that if you turn on one neuron, you will need less energy to turn on others. If enough neurons become activated, it no longer takes any energy at all to turn on additional neurons. The network then flips into a new phase of operation, able to store vastly more data than before. In a way, it behaves like a black hole. It is not literally a black hole, a cosmic sinkhole from which nothing can escape, but it has the key attributes of one, including an almost unimaginable internal complexity. "Black holes are the most complex systems that are possible," Dvali said.

The broader lesson is that all the diverse phenomena that physicists have discovered over the centuries may inspire AI and brain researchers. "We are good at dealing with systems with enormous numbers of degrees of freedom," Dvali said. "Biological systems are also systems with an enormous number of degrees of freedom. There is some expertise we can bring."

He and other physicists in this line of work have little patience for intellectual pigeonholes. They have found unexpected connections between material systems and artificial intelligence, and they see no reason to stop looking for more. Ultimately they want to reflect back on human intelligence. Physics is helping to explain neural networks, neural networks are like our own brains, and so physics is helping to explain us. In 1972 the physicist and philosopher Carl von Weizsäcker asked: "Are we strangers in the material world? Do we belong to it?"[69] Today's scientists answer him: No, we are not strangers. We belong. The principles that govern our intelligence are not so different from those that pack atoms into a crystal or matter into a black hole. Our brains call on many of the same self-organizing powers that are innate to matter. There is a continuity between inanimate and animate. As MIT's Dan Roberts said, "I see how to unify my way of thinking about intelligence, or myself, with thinking about the rest of the universe." In chapter 3, we will explore the theories of the human mind that are emerging from this disciplinary crossover.

3 | PHYSICS OF THE MIND

ON A MILD SEPTEMBER MORNING in 2018, I boarded the train in Thessaloniki and found I was seated next to Karl Friston. We were crossing Greece, going from one meeting on physics and neuroscience to another, and had a five-hour ride ahead of us.

A professor at University College London, Friston is the doyen of what is known as predictive coding or predictive processing, arguably the closest that neuroscientists have yet come to a grand unified theory of the mind.[1] His reputation is forbidding. "When he comes for a colloquium, the board fills with equations that take three days to unravel," I heard from Igor Aleksander, an AI pioneer at Imperial College London who is no mathematical slouch himself. Friston is one of the few academics these days who routinely wears a suit and tie, looking like a government minister who has no time for the little people. Yet I found him to be highly approachable and patient. I took the opportunity to learn not only about predictive coding but also about his own intellectual journey.

Though a psychiatrist by training, Friston told me he also considers himself a physicist. As a student he wanted to combine physics with neuroscience, but the discipline that does that—computational

neuroscience—hadn't yet been invented, so his academic coun-
selors suggested psychiatry. He chose a university program that
let him study physics on the side. "I could have this two-field approach," he said. "I wanted to understand the brain, but I had to
have this physics underneath. Physics has been in the background
even throughout my medical career." And he is completely on board
with the idea that physics needs the expertise of brain science if it is
to solve its own puzzles. "The anthropomorphic framing of physics
is accountable to neuroscience," he said.

Friston's daily routine harnesses a psychiatrist's awareness of how
brains work. Before bed, he reviews his notes on whatever problem
he is working on, so that his subconscious can turn it over in his
sleep. In the morning, he sits for two or three hours with his pipe,
no paper, no phone, just thinking. Confining the problem to his
working memory allows him to focus solely on its essence. Only
when he has a solution does he pick up a pencil. "You really have
to drill deep down to what you can solve," he said. Friston took up
this almost meditative practice when he was fourteen, he told me:
"It wasn't quite sex, but a buzz at clearing the decks: your problem
with no distractions." And he has been able to keep up the practice
even as a parent. Other working parents, who struggle to get five
minutes of alone time, might regard that as an achievement greater
than anything he's ever done in neuroscience.

The meeting we were heading to was intended to bring predictive coding together with another leading theory of the mind, integrated information theory. Competitors though they are, they have
much in common. Both theories are mathematically meaty, built on
simple principles, and grand in scope—in a word, physics-y. Both
view the brain as a special type of neural network and, like theories
of neural networks in general, are based as much on physics as on
biology. Both draw on physics concepts such as energy and causality.
Both look beyond the particulars of human biology to seek the qualities of consciousness in other animals, in machines, in collectives, in

inanimate matter—in anything, really. Both theories make debatable philosophical assumptions, but they align with most physicists' gut feeling that consciousness is a collective or emergent property.

Both, too, invert our usual conception of ourselves. It may seem to us that we have direct, unfiltered access to the wider world, but, in truth, each of us lives in a world of our own making. What we see and sense is a hallucination; it is actively generated by the brain. We recognize it as a hallucination only when it slips its leash—when the brain somehow fails to recalibrate our private world to match the evidence of our senses.

Both theories also have plenty of skeptics and could be utterly wrong. They are hardly the only theories of consciousness out there. Their claim to fame—that they find unity in diversity—is, to many, a flaw.[2] The brain is such an evolutionary palimpsest that trying to understanding it by going back to first principles could be a fool's errand. Advocates may also be guilty of confirmation bias in finding a role for their theories wherever they look. Experiments have been encouraging, but knockdown proof is hard to come by. We have no direct access to the conscious experience of anyone or anything other than ourselves, and even our self-knowledge is partial and often downright deceptive. So all our reasoning is ultimately by simile: I know I am conscious, and you are like me, so I presume you are conscious, too. Clearly, that gets sketchier for other animals, let alone computers.[3] They don't call it the hard problem of consciousness for nothing.

As long as we heed these disclaimers, though, we can explore the two theories with an open mind. Whether or not Friston and his colleagues succeed in their loftier ambitions, their ideas have already proved fruitful. These concepts have inspired algorithms of practical use in machine learning and fresh ways to help people with schizophrenia and locked-in syndrome. And, as later chapters will explore, these theories provide a foundation for exploring puzzles of physics that hinge on the nature of the mind.

WE SHAPE OUR OWN REALITY

Spare a thought for the frogs. Our modern understanding of the brain owes much to laboratory amphibians. Hermann von Helmholtz—one of the mid-1800s German "organic physicists" I mentioned in chapter 2—called them "martyrs of science."[4] If Helmholtz were alive today, physics, biology, and psychology departments would all fight to hire him, so important was he to all three fields.

Helmholtz found that a frog's calf muscles increased in temperature when they flexed, which got him thinking that energy is never lost or destroyed, but merely changes its form. From that he formulated one of the most important principles in physics, the law of conservation of energy.[5] Helmholtz was also interested in the workings of the nervous system. When he shocked one end of a nerve fiber in the muscle of a frog, it took a millisecond or two before the muscle contracted. (To measure such a short period of time, he built several ingenious contraptions. In one, when he applied the shock, electricity began to flow through a circuit, causing a magnet to begin rotating. When the muscle flexed, it activated a switch to shut off the power. The deflection of the magnet provided a record of how long the current had been flowing.) A millisecond sounds fast, but it was way longer than scientists had assumed, so the finding posed a new puzzle: How do our senses keep pace with the world? Why are we not always a beat behind?[6]

If the brain were a purely reactive system that took in sensory signals, processed them, and presented our conscious minds with sights and sounds, like a camera feed, we'd always be playing catch-up. The time that this processing takes varies, but a ballpark number is eighty milliseconds.[7] If our experience of the world lagged by that much, we could never play a fast guitar riff or hit a baseball pitch. And Helmholtz noted that our sensory information is not just perpetually late, but ambiguous—the same sensory data can be interpreted in multiple ways, yet the brain somehow chooses one interpretation.

Helmholtz concluded that the brain must be proactive,[8] that it does not start processing afresh every time we open our eyes or shift our gaze but gets a jump on things by drawing on past experience—"the unconscious processes of association of ideas going on in the dark background of our memory."[9] The brain tries to be ready for what comes. That doesn't always work; sometimes reality violates our expectations. When that happens, the brain corrects itself; it resets its expectations to get them right the next time.

Out of this insight would come predictive-coding theory, which goes further and says our expectations do not just help to shape our reality, they *are* our reality. We do not perceive the world as it is but as we expect it to be. Experience comes from within. "Every brain is literally a fantastical organ," Friston said. "It has fantasies that it uses to explain, in as parsimonious a way as possible, its sensorium."

Many of the quirks of perception make sense in this framework. For example, visual illusions often play off expectations. We might see lines or spots where there are none, because they are there in similar situations. We might not see an animal hidden in an image until someone points it out—and then we can't *not* see it. Helmholtz and later psychologists were especially struck by the phenomenon known as binocular rivalry, which sounds like a contest between bird-watchers, but refers to a conflict between the eyes.[10] Show a face to one eye and a house to the other, and you won't perceive some weird house-face blend. Instead, you will see the face, then the house, then the face again, alternating every few seconds. The explanation is that the brain knows full well that there's no such thing as a house-face, so it settles on one or the other. But because neither is a fully satisfactory representation of what the eyes are reporting, the brain vacillates.[11]

"Visual illusions, in themselves, are evidence for a working system of predictive coding," Friston told me. Sometimes, though, the brain fails to keep our inner life in alignment with the world outside. Friston, who treated people with schizophrenia during his medical residency, thinks that the condition occurs when the brain clings to its expectations so strongly that it overrules the senses.[12] According to him, when people hear voices, the problem isn't the hallucination per se, since all of us have been fooled about our sensations at one point or another. It is the failure to correct.

To be sure, schizophrenia is complex and can't be reduced to a single mechanism; still, it is striking to consider that the difference between psychosis and neurotypicality may be a matter of degree, not of kind. One might also wonder whether some of the observer or inside/outside problems we are having in theoretical physics occur because our mental models never quite mesh with reality. I'll return to that thought in chapter 5.

THE HELMHOLTZ MACHINE

Helmholtz was a major influence on Sigmund Freud, but by and large psychologists were skeptical of his idea that perception is a construct.[13] So the idea languished. Over half a century later, in the 1950s, the cyberneticist Donald MacKay proposed a similar scheme for AI systems.[14]

Up to that point, the early practitioners of AI had conceived of image recognition and other sensory processing as a one-way, feedforward process: input to output. MacKay suggested a special kind of feedback loop instead. An "imitative," or generative, module would make a prediction for the input, which a "comparator" would check against actual sense data before giving feedback. Through this process of self-critique, the system could learn to recognize patterns on its own rather than being coached by a human user. Not being merely reactive, but having an inner life, the system could be said to be conscious, MacKay suggested.[15] MacKay provided only the barest sketch of the design for such an AI, though. And his papers didn't cite Helmholtz—or anyone beyond his own immediate circle, really—so he may have rediscovered the principles independently.

Around the same time (and again failing to cite Helmholtz), communications engineers coined the term "predictive coding"— the very term neuroscientists (who are evidently more diligent about citing their predecessors) would adopt in the early '80s.[16] Predictive coding initially had nothing to do with brains. Instead, it addressed a very practical problem in telecommunications—how to efficiently transmit information—by taking advantage of human predictability. Speech and music, for example, comprise relatively few tones, which follow one another according to the rules of grammar, rhythm, and harmony. So you don't have to send an audio signal in its entirety; once you figure out the rules, you can transmit only the exceptions. The receiver will assume the signal follows the rules unless told other-wise. Apple Lossless music files and PNG image files use this sort of prediction to reduce file size, so you can download them faster.[17]

These ideas from psychology, physics, and engineering gradu-ally coalesced. In the early '80s Terry Sejnowski and Geoff Hinton fulfilled a crucial part of MacKay's vision with their Boltzmann ma-chine, which I described in chapter 2. As the first generative neural network, it could be set up to make predictions and update them in the light of new evidence. Unfortunately, training it was slow. Every time you fed in some data, you had to wait for it to settle into an internal equilibrium.

A decade later, Hinton and another group of coauthors came up with a faster machine.[18] They figured that if predictive coding could speed up communications, it could also speed up their net-work. Like feedforward networks that learned to recognize puppy and kitten pictures, theirs had multiple layers to perform multiple stages of processing, first looking for pixel-to-pixel differences, then identifying larger geometric patterns. But unlike those networks, it sent signals in two directions. Input signals entered at the bottom layer and percolated upward, while predictions trickled down—so the same network did both recognition and generation. The two functions alternated. During what the researchers called the "wake" phase, signals flowed upward and the network analyzed an image—say, a puppy or a kitten—for its geometric structure. During the "sleep" phase, signals flowed downward and the network generated an archetypal image of whichever animal it had in mind. In other words, it dreamt. And its nocturnal visions primed the recognition system for another pass.

The researchers called their creation the Helmholtz machine, which was a double entendre. Not only did their network generate predictions, as Helmholtz had suggested, but it was also trained using a seemingly unrelated idea from Helmholtz's work in physics: free energy. As I mentioned in the previous chapter, the idea of using energy to describe neural networks was one of the most important contributions physicists made to the field. Like anything in the physical universe, a network seeks out its lowest possible energy. For the Helmholtz machine, the relevant type of energy is known as free energy, which represents the portion of a system's total energy that is useful for creating a lasting change. A system can exhaust its free energy, at which point it settles into equilibrium.

For machines, gases, and other material systems in physics, free energy can be calculated from measurable quantities such as temperature. When applied to neural networks, free energy has a more abstract meaning related to information processing: it quantifies the precision-generality trade-off. An accurate but complex network has a high free energy, signaling that it is failing to discern the underlying regularity of the data. A simple but error-prone one also has a high free energy, signaling that it is just plain bad. Somewhere in between, the network finds its lowest free energy and strikes a balance between complexity and verisimilitude. Its predictions will be good, but no better than they need to be.

By comparing the wake and sleep phases of the network, you can calculate how to reduce the free energy without having to wait for the system to reach equilibrium. And that makes the Helmholtz machine much faster than the Boltzmann machine. The elegance of the Helmholtz machine got Friston hooked on the importance of predictions. "Hinton was probably the most prominent figure in my recent intellectual mentorship," he told me.

THINKERS VS. BEETLES

These early efforts—MacKay's self-critiquing AI and the Helmholtz machine—introduced the principle that signals move both up and down the hierarchy of processing. Raw perceptual input comes in from below, while expectations from past experience trickle down from above. So far these models were only loosely connected to neurobiology. In 1999 the computer scientists Rajesh Rao of the Salk Institute and Dana Ballard of the University of Rochester took the next step by using a multilayered predictive network to model one of the brain's most impressive feats: vision.[19] In so doing, they clarified how the upward- and downward-moving signals mesh.

Suppose you are scanning a crowd for your friend's face. Having formed the expectation of seeing his face, the top layer of your visual system makes a prediction that the layer beneath it will detect face-y shapes, such as a pair of symmetrically placed ovals (otherwise known as eyes). That layer, in turn, predicts that the next one down will find the lines, curves, and angles that make up those shapes. Rao and Ballard's model had two layers, but if there were more, they'd break the geometric figures down further still.

If a layer doesn't see the predicted geometric forms, it marks the prediction as erroneous and sends a signal up the chain, perhaps forcing the uppermost layer to predict something besides faces.

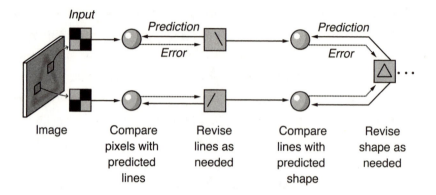

3.1. PREDICTIVE CODING. Consider the brain's perceptual system as a processing pipeline that takes in data from the left, passes it through various stages of analysis, and produces an interpretation on the right. In this example, it dissects an image into lines and then a triangle. What is novel about predictive coding is that there is also a counterflow of information. The brain forms high-level expectations on the right, passes them toward the left, and updates those expectations if they fail to match reality. The brain expects to see a triangle, which translates into lines of various orientations, which translates into patterns of light and dark. If, try as it might, the brain can't find lines, it readjusts to scan the image data for other geometric figures. The two-way flow speeds up processing and resolves ambiguous interpretations.

Eventually the network hits on a prediction that matches the data. The two-way flow of information combines the raw data with contextual cues. It's like assembling a jigsaw puzzle (without peeking at the answer) by working on both large and small scales. You reckon there should be a face, so you go looking for eyeballs and nostrils, but you don't find any. You do notice a lot of light blue pieces, though. So now you think there must be sky and begin checking whether the blue pieces fit together. Back and forth you go. The big picture clarifies the placement of individual pieces, and the fitting together of pieces confirms that you're not just making it all up.

Taking a page out of the communications engineers' book, the upward-flowing signals are not complete, raw data, but just the errors—what wasn't predicted. There's no sense in clogging up the brain with data it expected—it should save itself for surprises. In Rao and Ballard's network, errors play a dual role. For the immediate task of image recognition, errors prompt the network to try a different mix of geometric figures. Over the longer term, the remaining errors prod the network to rewire itself to improve its accuracy. "The extra thing that the Rao and Ballard paper brought to the table was that the same prediction errors that drove inference were also responsible for learning," Friston explained.

This network operated on snapshots: static images as opposed to video feeds or other inputs that change over time. Its "predictions" were thus applicable only to one particular instant—the system couldn't hazard a guess about what might happen in the future. Around the same time, the electrical engineering professor Jun Tani, now at the Okinawa Institute of Science and Technology, was beginning to tackle the time dimension. He approached the problem from a different angle altogether: robotics.

Roboticists in those days were split into two camps. Some designed their mechanical creatures to be thinkers: look around, formulate a plan, make it happen. Others made beetles: scuttle about, follow trails, rush toward lights. Shakey, a pioneering robot in the '60s, was a thinker. It had a central processor to analyze video images, obey typewritten commands, and create maps and plans. It was also painfully slow and bumbling—its creators called it "Shakey" for a reason.[20] The original Roomba vacuum cleaner robot is a beetle. It reacts to its surrounding environment without any sort of plan, which makes it nimble, but also blundering; you're endlessly having to rescue it from door thresholds and carpet tassels.

Tani argued that it didn't have to be an either/or situation.[21] He showed that a control network could combine a thinker layer and a beetle layer: thinker plans, beetle executes, thinker rethinks. The system has the same two-way signal traffic as predictive coding, although Tani didn't make that connection for another decade. In 2003 Tani built a crucial feature of our brains into his network:

the ability to make predictions on multiple scales.[22] As in earlier designs, higher layers took in broader patterns, while lower layers focused on details. But now these were patterns not just in space, but also in time. Higher layers looked further ahead; their predictions were judged against sensory input arriving a few moments later. "The higher levels are allowed to activate only slowly; the lower are much faster," Tani told me when I visited him in 2017. In addition to making sense of a changing environment, the system could break down its own movements into a sequence of motor commands. Higher layers formulated the general plan of, say, throwing a ball, and lower layers figured out how to bend back the arm, snap it forward, and let go of the ball at the right moment. This system finally proved Helmholtz's point: to function in the real world, we have to stay slightly ahead of the action.

LEARNING ALL THE TIME

Friston and I were so caught up in our conversation about the history of predictive-coding theory that it took us a while to register that a strange drama was playing out on our train. People boarded, took their reserved seats, got into an argument, and ended up hauling their luggage elsewhere in the train. At the next station stop, the same—so it went for hours. Not knowing Greek, we were mystified. Then we saw why there was trouble: the seat numbering on this train was really weird. It jumped around, and people who thought they'd reserved adjacent seats found themselves rows apart. Friston and I never did figure out the logic behind the seat numbers. Sometimes our brain can't make sense of the world around it and must find a way to live with ignorance.

One of the many insights of predictive coding is that to live is to learn. The brain is learning every waking moment, not just when we're cramming for a history test. Whenever we see, hear, or feel something we weren't anticipating, the brain adds this novelty to its store of knowledge. But how does it take away the right lessons? That's the trick. Sometimes things happen for a reason, and sometimes they just happen. The brain has to try to tell the difference, or else it will see meaning where there is none. "You want to attenuate fake news," Friston said.

According to the theory that he and others have been developing over the past twenty years, the brain has several ways to penetrate the fog. First, the brain is cautious in assimilating new information. It does not completely give up what it knew before, only nudges it. Through gradual self-correction, it comes close to the mathematically optimal way of keeping your knowledge up to date, called Bayesian inference after the eighteenth-century English mathematician Thomas Bayes. (To employ this procedure consciously, you express your certainty about your knowledge probabilistically and, using a simple formula, update the probabilities whenever you come across new information that has bearing on your belief. This works not just for perception but for any complex situation where you have to pull together information from multiple sources. It is a healthy intellectual attitude to think in terms of probability rather than absolute certainty. The Bayesian approach also forces you to articulate your prior beliefs; none of us comes to any new situation entirely free of bias.)

Second, the brain assigns a degree of confidence, or precision, to each of its predictions. High precision means a greater sensitivity to input. If the brain makes an imprecise prediction and it's wrong, so what? It didn't expect to get it right. But a precise prediction that's wrong is another story. The brain does the equivalent of saying "My bad," and revamps the model. In concrete terms, Friston said, the brain encodes precision in the concentration of neurotransmitters, such as dopamine.

Third, the brain adjusts the precision of its predictions as it climbs the learning curve.[23] When starting to learn something, it cranks up the precision so that a new model quickly takes shape. At some point, though, continued refinements bring diminishing returns, so the brain dials back the precision and tolerates the remaining discrepancies as random scatter. In other words, it avoids overfitting the data. Friston noted that this process is very similar to the procedure of "annealing" quantum neural networks that I described in chapter 2. The network is initially very plastic, the better to absorb new data, but gradually solidifies. These adjustments happen automatically in our brain, but we experience them viscerally. We feel the thrill of novelty, then the satisfaction of mastery, and finally restlessness, pulling us toward new challenges. In this way, our brain keeps itself on the cusp between frustration and boredom.[24]

Within this framework, there is a range of different learning styles. Friston and others have suggested that autism may be a tilt toward high precision, leading to a kind of perceptual perfectionism.[25] People with the condition are quick to pick up on details and changes, with the trade-off that they often have trouble seeing broader patterns and tuning out background noise. Other features of autism, such as a preference for routine, may be secondary—they could be ways autistic people adjust to their heightened sensitivity to novelty.

DOING IS SEEING

If predictive coding were just a mechanism for perception, that would have been enough. But Friston—searching, like all physicists, for unity—realized in 2003 that the same mechanism could drive movement, too.[26] Faced with a discrepancy between its model and the world, the brain normally updates its model, but it has another option: update the world. By flexing a muscle, it can make the world match the model.

Suppose you want to pick up your coffee cup. The brain starts by predicting that you will do so. This expectation filters down through the neural network to lower levels, which break it down into the specific sensations you should feel in your body. The predictions reach the reflex arcs that reside in the spinal cord and control your muscles directly. Your arm rises and moves forward, your joints creak, these sensations propagate back up the nervous system, and the brain goes, Yup, that's just what I was after. "Actions are making your predictions come true," Friston said.

The principle of repurposing the perceptual system for motor control goes back to the nineteenth century.[27] It has the huge advantage that the brain does not need to plot out each bodily motion. The body has a flexibility that any robot would envy, and physiologists realized that controlling our musculature centrally would be a mammoth computational challenge.[28] So the brain takes advantage of the body's natural kinesthesia. With minimal prompting, the body's mechanical linkages can execute motions on their own. On this account, even the practiced movements of a dancer or a saxophonist are no different in kind from a knee jerk in response to a doctor's hammer.

According to Friston's way of thinking, which he calls active inference, the brain is not the body's helmsman or puppeteer, but its dreamer. Brain and body are bound up in a mutual project to predict the world successfully. Sometimes the brain does the work, sometimes the body. "The way that we move—the intention to move, the willed actions that we enjoy—are actually prior fantasies that I am going to do this," Friston said. "The body—the reflexes, the muscles—realizes those fantasies."

How do the two divide their labor? The concept of predictive precision is the key. To initiate movement, the brain forms a belief that it is already moving. The belief is false, a kind of deliberate hallucination. But the brain assigns this belief a low precision such that it lingers, unchecked. Having stopped itself from correcting its own false belief, the brain passes the buck to the body. The reflex arcs, sensing the mismatch between predicted sensations and actual sensations, spring into action. The brain is "letting the body do the work in terms of righting the prediction error," Friston said. The prediction becomes self-fulfilling, but only because the brain temporarily blinkers itself.

In this way, our every action requires a suspension of disbelief. To do well, you need to stop worrying about doing well. When athletes choke, the problem is often hyper-self-awareness, a pathological perceptual honesty. Friston has argued that people with Parkinson's are stiff and slow because their brains can no longer adjust their dopamine levels.[29] They thus lose the ability to tell themselves the little lies that are needed to get anything done. Their brains predict that they're moving but immediately realize that the prediction is wrong rather than overlook the truth and let the body act.

Predictive-coding theorists think our emotions and our sense of self are rooted in this brain-body dynamic. "My perception of being a self is completely bound up in my brain perceiving and interacting with physiology," Anil Seth, a neuroscientist at the University of Sussex who specializes in how the brain predicts internal bodily states, told me. "Understanding conscious selfhood starts from understanding predictive models of the control of the body."

WHY STOP WITH perception and motor control? Over the years, Friston has steadily broadened the scope of predictive-coding theory, and he now thinks that almost every aspect of the brain, and indeed of life in general, can be traced to the making and refining of predictions. Let's set aside the puzzles of natural and artificial intelligence for a moment and ask a more basic question. How does anything endure? Nature, red in tooth and claw, slaughters the unwary; the world is a ceaseless churn that eventually wrecks everything within it. Any sustained structure must stabilize itself against external insults. Living things have this ability; arguably, that is their defining characteristic, although lots of systems that we do not traditionally consider living are able to sustain themselves, too, such as hurricanes. What makes this endurance so remarkable is that these are open systems. Our bodies cannot wall themselves off because the world that would destroy us also sustains us. We eat, excrete, breathe in, and breathe out, regularly exchanging new molecules for old ones. Yet we persist.

If a living thing is to survive these depredations, it must be ready for them, so it can adapt. Most inanimate things don't do that, and sooner or later they disintegrate, so by default whatever is left follows this principle: its internal states mirror and anticipate those of the outside world. "There is an implicit model of the world encoded by its internal states," Friston told an audience at one conference I attended.[30] To describe this modeling process, Friston repurposed the concept of free energy that Hinton and his coauthors introduced to train neural networks. In this context, free energy quantifies the cost-benefit trade-offs that every living thing faces. Hunting for food is risky but has benefits if it succeeds, and an organism's internal model settles on a course of action that weighs the pluses and the minuses. The model itself entails a trade-off. An elaborate model is better at navigating the world, but takes energy to create and maintain. In both cases, an organism strikes the optimal balance when its free energy is minimized.

If an organism or self-sustaining nonbiological system models the world, it has a kind of mind—rudimentary, perhaps, and not always something we would recognize as a mind, but nonetheless some kind of inner mental life. Without awareness of its surroundings and of itself, a thing could not act to ensure its own survival. Friston and his coauthors thus place sentient beings on a continuum with living things in general and with inanimate matter.[31] A rock or snowflake might not have a lifelike mechanism to stabilize itself, but some nonbiological systems do. In 2020 Friston and several coauthors gave the example of Earth's climate, which has been stabilized by various feedback loops operating over geologic time. For instance, the sun has gradually gotten brighter, yet Earth has remained fairly hospitable, in part because plants and plankton are sensitive to temperature and, through their activity, alter the atmospheric composition to offset the more intense solar radiation. (This is all separate from climate change today, which is happening much faster and for different reasons.)[32] The researchers suggested that plants and plankton can be thought of as modeling the trend in solar radiation.[33] Other physicists have been thinking similarly expansively[34]—seeking, as David Wolpert at the Santa Fe Institute put it to me, a "deeper and more physics-based notion of living systems, rather than just organic chemistry."

HOW ERROR MAKES US AWARE

As for the deeper puzzles of consciousness, such as why we have subjective experience at all, the picture gets much hazier. Of the predictive-coding theorists I've talked to, Jun Tani had the most evocative ideas.

Unlike many academics, Tani did not tell me of a precocious youth spent contemplating cosmic mysteries or winning math olympiads. On the contrary, his childhood in Japan was a struggle. "Language came very late for me—maybe three years old," he told me. "In my elementary school, I was . . . always the worst in my class. . . . In my school days, I was so slow. Now I'm reflecting back why."

He became an industrial piping engineer. His job: stop pipes from banging. The need for better plumbing took him down some unexpected rabbit holes. As he thought about fluid flow patterns, it occurred to him that consciousness, too, is a search for patterns. That insight, along with a desire to understand his own learning challenges, inspired him to study cognitive science. He went to the University of Michigan for master's degrees in mechanical and electrical engineering in the mid-1980s and, on returning to Japan, established himself as a leader in using robots as models for neuroscience. Robots face the same control challenge as the brain: responding in real time, with limited computing power, to a capricious environment. After all, our brains evolved not to do crosswords or cogitate on philosophy, but to solve the practical problems of survival and reproduction.

In 1998 Tani argued that embodiment is what drives subjective experience.[35] When our brain makes a prediction and it proves right, sensory inputs do not rise to the level of conscious awareness; we are on autopilot. But when the prediction is wrong and not easily fixed, the discrepancy summons our attention so that the brain can bring everything it has—its senses, its knowledge, its powers of reason—to bear on the new situation. "That is why, phenomenologically, we feel consciousness," Tani said. In short, we are only aware of thwarted expectations. And there is a lot of thwarting. Nothing in the real world is ideal, nothing ever goes according to plan, so the brain is always erring and recalibrating. If not for these continual imperfections, we'd have no need to be conscious.

As an example of how physicality creates our mental experience, Tani, who has played the contrabass in jazz ensembles since college, talked about musical improvisation. Jazz, more than most musical genres, mythologizes the mistakes that come from testing your limits.[36] Tani can play a basic melody line automatically, but when his turn to solo comes, he is pushed to try something new. He intends to do one thing, but another thing happens, and he goes with it. He feels entirely present. "You want to make a high note now, but it's impossible," he said. "Therefore you struggle, and that kind of struggle, you're very conscious about it. After that struggling, some really new pattern, tune, or tone comes out. . . . Without embodiment, you don't get stuck. If you can generate whatever sound very freely, there is no way of creating things."

New perceptions are endlessly churning through our sensory system, tickling our conscious awareness. Even a seemingly static scene is capable of causing surprise. Tani imagined a glass of red wine. "You have an expectation of the redness of the wine, but you have the reality," he said. "In the real one, a particular Chianti in the glass, you see the red is so deep, and it's different from your expectation." Some new quality of color and light catches your eye; your perception never stabilizes. "A small error always remains, because this is the real world. It is a bit different from what you think. That is qualia, I feel. Reality is never able to be conceived of perfectly."

INTEGRATED INFORMATION THEORY

When Friston and I arrived in Athens, we and the other scientists and philosophers took a ferry to our hotel. Over the next two days, Friston swapped ideas with advocates of integrated information theory (IIT), the other theory of the mind that captivates many physicists these days. IIT doesn't seek a grand unified theory of brain function as predictive coding does, but zeroes in on consciousness. Whereas predictive coding takes our ability to keep pace with the world as its starting point, IIT begins with another core feature of the mind: its unity.

There's no jagged crack running down the middle of our vision. We don't see the left side detach from the right, or an object's color float free of its shape. It all coheres. Our sensations form a single field of experience. "Having two experiences—it would mean you are now two people," said Giulio Tononi, a neuroscientist at the University of Wisconsin–Madison and the father of IIT. "That's inconceivable. *This* is my experience. It doesn't even make sense to say now my experience is split in two. Mine and that of whom?"

Descartes took the unity of perception, and of the mind generally, to be highly significant. In 1641 he wrote: "I can distinguish no parts in myself but understand myself to be a thing that is entirely one and complete. . . . It is one and the same mind that wills, that senses, and that understands."[37] And if the mind can't be decomposed into parts, he figured, it can't be explained at all, at least not in the same terms as material systems. This was his articulation of the hard problem that so troubles today's neuroscientists and philosophers of mind.

Tononi suggests that the unity of experience reflects the unity of brain activity: When different parts of the brain work together in harmony, we have experiences. When brain regions disconnect from one another, as in a deep and dreamless sleep, we are effectively automatons. Consciousness, he thinks, does not require any specialized module or processing hierarchy in the brain. It is a natural product of wiring neurons together so that the system becomes more than the sum of its parts. On that intuition, Tononi and his colleagues have constructed a complicated and, at times, esoteric mathematical theory. As I'll elaborate on in a moment, IIT quantifies the degree of unity, pinpoints the participating structures in the brain, and relates their patterns of activity to subjective experience. Notably, it rises above the neuronal nitty-gritty. The theory holds that any system that shows the same unity as our own brains will be conscious in the same way. Octopuses, plants, robots, and ant colonies might, depending on their internal arrangements, have some modicum of consciousness.

This generality is a strong selling point. "I admire IIT in that it doesn't presuppose that consciousness has to be only biological," said Ron Chrisley, a cognitive scientist at the University of Sussex and cofounder of the AI startup Tenyx. "So it at least allows for the possibility we could encounter another species, or there could be artificial agents that are not biological, or maybe groups of entities that are not themselves biological. What I like about IIT is that it doesn't prejudice the issue from the outset by defining consciousness to be a biological condition."

I first met Tononi at a conference of the Foundational Questions Institute in Puerto Rico in 2014. There, he introduced his theory to a crowd of mostly physicists and cosmologists, and it made a big splash. Many took up the gauntlet, some with the aim of refining it, others with the intention of criticizing it. Tononi's theory offers an irresistible challenge to physicists—it has its share of gaps and fuzziness, which is a good thing, because nothing motivates a scientist more than seeing a problem and having the urge to fix it.

Plausible though his basic thesis about unity may be, it is still a leap of faith. Skeptics dispute whether conscious experience is truly unified, or whether it requires the brain to be.[38] To give a real justification for why the structure of experience has to mirror the structure of the physical brain, you'd have to solve the hard problem, and one strength of IIT is that it sets that problem aside. It seeks to actually do something to quantify and describe consciousness, rather than go round in philosophical circles. You can think of IIT as a controlled way of extrapolating from our own experience to other systems. It starts from the fact that we are conscious, proposes unity as a possibly relevant feature, and looks for the same property elsewhere. Every theory has to start somewhere, after all.

"KNOCK ON THE BRAIN"

Tononi models the brain as a neural network. Like all such networks, it consists of basic units such as little switches that are wired together and toggle on or off in response to other devices. For purposes of studying consciousness, he is interested in the network's internal activity rather than its inputs and outputs. The network is typically so complicated that you can't just look at it to see how unified it is; it might even be a cluster of disconnected smaller networks. So Tononi conducts a thought experiment to ascertain its unity: Split the network in two, manually turn the neurons on one side off and on at random, and see what happens to the other. The bigger the effect, the more tightly the two sides must be integrated. There are multiple possible dividing lines along which to split the network into two sides, and you try them all. On the principle that a chain is only as strong as its weakest link, the dividing line that yields the least effect (the least integration) defines the overall integration of the system.

To get a sense of what this split-and-randomize thought experiment is calculating, imagine that you are at a banquet with what looks like a big table covered with a tablecloth, and you wonder whether it is really a single big table or a cluster of small tables that are pushed together. You kick one leg of the table and see what happens to the plates and glassware. If they're on the same table, they'll probably shake in response—this is like an integrated network. But if they're on separate tables, they may be entirely unaffected, which is like a disconnected network.

Tononi's research team has an online app that will perform this thought experiment for a network that you specify. The outcome is a quantity that Tononi denotes by the capital Greek letter phi, Φ, which represents the amount of information that is held collectively by the network rather than stored in its individual elements. There's a Goldilocks effect at work: the network should be neither too loose nor too tightly bound. In a cluster of disconnected networks, information remains atomized, so Φ is low. But a fully connected network, where every unit is linked to every other, behaves as a monolith; the network has very little information-storage capacity, so Φ is low again. Somewhere between these extremes, the network stores the most information collectively and Φ reaches its peak value.

Defining Φ is one thing, but calculating it is another. You have to go through all the possible ways to cleave the network in two, of which there are a gargantuan number. The online app gives up if the network has more than eight units.

But Tononi's recipe gives physicists something to sink their teeth

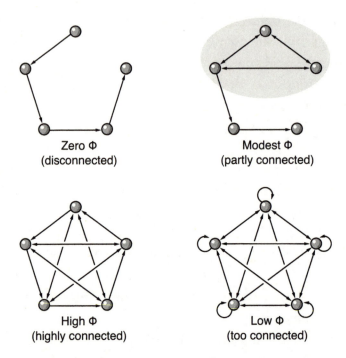

Zero Φ
(disconnected)

Modest Φ
(partly connected)

High Φ
(highly connected)

Low Φ
(too connected)

3.2. INTEGRATED INFORMATION THEORY. Integrated information theory supposes that a neural network is conscious if its units are wired together tightly enough that they develop interesting group dynamics. The theory lays out a procedure to quantify their degree of integration, denoted by Φ. A disconnected, or feedforward, network, without any closed loops, has zero Φ. If some units are tightly integrated while others are feedforward, only the tightly integrated section is conscious. With enough wiring, Φ peaks. Beyond this point, additional wiring actually reduces Φ, since the network acts as a single block rather than as a true network with internal complexity.

into. It introduces concepts that are useful far beyond their original context, providing a way to quantify the complexity of any system, conscious or not.[39] When a network maximizes its Φ, it is poised at the critical point, at the edge of chaos in the way I talked about in chapter 2 in reference to optimizing neural networks.[40] There, it exhibits its most complex behavior. This condition, called criticality, also describes materials making a transition between two states of matter (one of my favorite examples is known as critical opalescence, a liminal state between liquid and vapor that looks like a dense fog). Based on such parallels, physicists have been able to help Tononi by offering more tractable ways to perform the calculation, or at least approximate it.[41]

Tononi equates Φ with the degree of consciousness of a neural network, be it an animal brain, an AI system, or really any network at all. This rubric appeals to our intuition that consciousness is not simply present or absent; it exists on a spectrum. You can be anywhere from comatose to wide awake and, even when awake, from zoning out to being fully present. Some achieve a heightened consciousness through meditation. We allow that other mammals and living things might be conscious, but assume, perhaps parochially, that they are conscious to a lesser degree than we are. IIT attributes these differences to the degree of brain integration. Placing all these conscious states on the same spectrum is probably oversimplifying the matter,[42] but Tononi and his colleagues have some experimental justification for it: they have performed the split-and-randomize thought experiment on actual brains to measure, if not quite Φ, then something close to it. Put simply, they have built a consciousness meter.

"We wanted a tool to perturb the brain—knock on the brain and see whether it was responding," Tononi told me. Their device combines two standard neurological instruments: a transcranial magnetic stimulation device, which is a handheld magnetic coil that zaps the brain to treat conditions such as depression;[43] and an electroencephalogram (EEG) skullcap laced with electrodes to eavesdrop on the brain's electrical signals. The coil delivers a magnetic pulse that acts like a clapper striking a bell, and then the EEG registers the resulting brain waves. An awake person's whole brain reverberates, indicating a highly interconnected neural network. But in deeply sleeping or anesthetized people, or people in a persistent vegetative state, the response is localized and muted.[44] The response indicates whether someone is conscious even if they're unresponsive. (This is still an experimental procedure, and these people or their legal surrogates must give informed written consent.) They've also tested the system on rats and fruit flies.[45]

Having your brain zapped to probe your consciousness sounds trippy. Alas, having tried it myself, I can report that you don't feel much of anything. When the doctor administering the test placed the coil against my head, the placement was a little bit off, and my finger twitched because the system was stimulating the motor areas of my brain. For a moment, I felt like a marionette. When properly situated, though, the system didn't affect any brain functions. "It should prove the state of consciousness, without altering it. . . . There should be no induced hallucination or anything like that," said Bjørn Erik Juel, a postdoc in Tononi's lab. Afterward, I looked at the electrical traces and saw the waves of activity across my brain. They revealed nothing of what I was thinking, but did prove that different parts of my brain were communicating with one another and, if you accept IIT, that I was conscious.

Calculating Φ is just the start. Not only can you ascertain whether a network is conscious, you can also hazard a guess as to what it is conscious of. A thought, a feeling, a memory: these are configurations of the network. If you map its wiring in detail and work out how the neurons are switching on or off over time, you can read its mind. The brain-zapping procedure I experienced was too crude to look into my thoughts, but Tononi and his colleagues have done this for simple artificial networks. Each possible pattern of activity corresponds to an experience, and in some cases they can work out the structure of that experience—whether it has the features we associate with, say, color or spatial relations, as I will explore in chapter 8.[46] But whether the experience is truly comparable to red or spatiality—or anything else we can identify with—is beyond mathematical analysis; only the network itself knows. The key point is that researchers have found a way to use mathematical and experimental methods to describe a correlate of experience.

THE MAN WHO IMPALED HIS BRAIN

Which part of your body is *you*—your conscious self? The brain has multiple parts and levels of organization, and for each, you can find researchers who think it's the site of consciousness. Some put it in the cerebral cortex, some in the brain stem. Some think consciousness is not a property strictly of the brain, but involves the body, environment, and culture. Others go to the other extreme of scale and place it at the subcellular level, inside neurons.

IIT provides a clear path to answering the question: calculate Φ for the various candidate sites and look for the one with the highest value. That is hard to do in practice, but Tononi takes an educated guess. In humans, the cortex—especially its rear, or posterior, portion—is richly structured and interconnected, so he predicts that consciousness resides there. "Grid-like areas like the ones in the back [of the cerebral cortex], and even more so a pyramid of grids, are a fantastically good substrate for phi," Tononi told me.

He and his colleagues cite neuroscience data in support of this supposition. Mélanie Boly, a neuroscientist who is also at Wisconsin, spoke in a lecture of a college student who, during the Spanish Civil War, escaped out a window, slipped, and impaled his head on an iron spike.[47] The front of his brain was mangled; it's hard to imagine a more devastating injury. Yet he survived and recovered. He got married, had two kids, and led a fairly normal life. The only indication that something was awry was that he was absent-minded: he had trouble sticking to a task and told the same jokes over and over. I know many people who never had an iron rod through their skull who are like that. But an injury to the rear of the brain leaves people much worse off, Boly said. They can lose entire categories of experience, such as color, or sink into a vegetative state from which they never emerge.

"Prefrontal cortex, which is a huge and highly evolved part of the brain, doesn't seem to matter too much," Tononi said. "But some parts in the back of the brain, for instance, in the back of the cortex—touch those, and you're going to screw up something with consciousness." As further evidence locating consciousness at the back of the brain, Jun Kitazono at the University of Tokyo and his colleagues analyzed the brain activity of two macaque monkeys, measured using implanted electrodes. When the monkeys were awake, the activity in the back of their brain was integrated; under anesthesia, it became fragmented.[48] To be sure, skeptics have pushed back, and the locus of consciousness remains an open question.[49]

One aspect of this question is how even to define the various structures. There are no perfectly sharp boundaries in nature. Where does the brain give way to the rest of the nervous system? What are the limits of the body? "Physics doesn't really tell us what are the objects, the entities, in the world," Tononi said. "It's all a giant field of things; it's very complicated. But it doesn't really put borders in any fundamental sense. So how do we know where things end and begin?"

IIT offers a principled way to carve nature at its joints because it analyzes any network at all. Typically that's the brain or a part of it, but you could equally well consider the extended network formed by the brain, body, and outside environment, or the miniature network of proteins within cells. Within each network, the theory will identify which sections are the most internally coherent, and by comparing the values of Φ at different scales, you can pinpoint which scale is relevant. Therefore, although everything is connected to everything else and boundaries are fuzzy, you can still delineate the sentient portion of your brain or any other system.

BOTH PREDICTIVE CODING and IIT imply that minds are everywhere. Any time you come across a structure that sustains itself in a chaotic environment or has a high degree of integration, you are looking at a thing with the potential for inner experience. Both theories thus embrace a version of panpsychism.[50] To be sure, these theories are very different from traditional panpsychism, because they place consciousness firmly within physical science, not beyond it, and do not claim that consciousness is a new fundamental property of nature.[51]

Both theories also pull back from saying that absolutely everything can be conscious.[52] Without such limits, panpsychism can get out of hand. Consider the multiple levels of organization in the nervous system. Each is a thing in its own right, with some degree of internal integration. Does that mean that our posterior cortex *and* both brain hemispheres *and* the areas within them *and* the neural circuits that make them up *and* all our billions of neurons—all the pieces of us that do not strictly have a Φ of 0—are individually conscious? If your entire cerebrum has a certain value of Φ, your cerebrum minus one neuron also has some value of Φ, as does your cerebrum minus two neurons, and so on. That's an awful lot of minds in one head.

The idea that our heads are stuffed full of conjoined minds not only is unsettling, but also threatens to make IIT unfalsifiable. All those minds would have different experiences, so the theory could never explain why you have the experience you do. Your brain might allow you to see in color, but a subset of its neurons might have no concept of color. It would just be the luck of the draw that your consciousness arises from one subset rather than the other. "It would mean that your experience could be anything," Tononi said.

To eliminate this ambiguity, he argued that there is only one physical system—the brain—so it can't support an infinite tower of minds. If you consider all those possible structures that might conceivably be identified with consciousness and calculate their Φ values, one of them will come out on top, and that's where consciousness resides. Two minds can't share the same neurons at once; the mind that is more tightly integrated prevents the other from existing at all. "Only one entity can exist," Tononi told me. "There cannot be any overlap whatsoever."

That said, several consciousnesses might arise within the brain as long as they don't overlap with one another.[53] Tononi has speculated that when your mind drifts on a long drive, for instance, it may just be your "main mind" that drifts. A temporary minimind may form elsewhere in your neural circuits, literally an autopilot that remains conscious of the road and keeps control of the car.[54] Other researchers, too, have suggested there could be multiple "islands of awareness" in the brain, some of which might be disconnected from all sensory input and left to think in utter isolation.[55]

The no-overlap rule not only stops minds from proliferating endlessly, but also provides a way to meld them together, thus solving an old difficulty for panpsychism known as the combination problem.[56] If two independent minds with a certain value of Φ link up to form a single mind with a higher value, they will cease to exist as individuals and assimilate into the collective. Indeed, this happens to us every morning when we wake up: individual regions within the brain, which might have some modicum of consciousness on their own, fuse together, and gradually the brain as a whole regains consciousness. Something similar might well happen in ant colonies.[57] For groups of humans, though, it's unlikely. Even our most tight-knit group is less integrated than a single brain, so human consciousness resides at the individual level. A corporation is not a conscious person deserving of the same rights as actual people, whatever the US Supreme Court may say.[58]

TO DO IS TO BE

In the second half of the twentieth century, the dominant theory of the mind was functionalism, the idea that mind is as mind does.[59] In one popular version, the mind is the brain's software, its main function is to process information, and the details of the hardware don't really matter; any two neural configurations that perform the same functions will be equally conscious. IIT, on the other hand, focuses on the unity of both experience and the corresponding brain activity, so it identifies consciousness with the structure of a system rather than with its function. "It's not what it does," Tononi said. "It is what it *is*."

Consider two networks. They do exactly the same thing—the same inputs yield the same outputs—and thus are indistinguishable from the outside. But when you open them up, they look very different. One is a neatly arrayed feedforward network, in which input produces output and stops there. The other is a feedback network in which signals can loop around. (These are like the top left and bottom left networks in figure 3.2.) The first network is not integrated. Its later stages of processing depend on the earlier ones, but the earlier ones are independent of the later ones, so there's no potential for collective behavior. IIT deems it unconscious. It is what philosophers call a zombie—a body without a mind.[60] The second is highly integrated. Each component depends on every other component. The network is capable of new and unexpected behavior that might surprise even its designer. IIT regards it as sentient.

Feedback is the mechanism that produces the unity that IIT identifies with consciousness. Its importance is a recurring theme in AI, neuroscience, and physics. Hermann von Helmholtz, John Hopfield, and other pioneers thought that feedback was essential to the creation of a mind. Many theories of consciousness other than IIT, not least predictive coding, also stress it. Not only is feedback more interesting than a feedforward architecture, but also it has practical advantages. It is usually more flexible and allows each unit to perform multiple roles. "In order [for a feedforward system] to be functionally equivalent to a system that has feedback, you need many more units and connections," Tononi said. Those efficiencies matter a lot to an organism competing to survive in a world of scarce resources and may help to explain the origin of consciousness in the history of life on Earth.[61] "It suggests a reason why consciousness, if it is integrated information, might have evolved," he told me.

The distinction between the function and the structure of a network also bears on the question of what would happen if you simulated your brain on a computer. Suppose you built a chatbot from this simulation. The machine would make the same jokes you do, recommend the same music, and send the same racy texts to your lover—so everyone would think it's really you. But would there be any feeling behind its utterances, or would they be delivered purely mechanically? Would the machine be conscious? Science-fiction shows such as *Black Mirror* and *Westworld*—which, at the rate technology is advancing, won't be fiction for much longer—assume that it would, but cognitive scientists and philosophers are divided. Tononi sides with those who say: not so fast.[62]

He told me his colleagues once designed a tiny programmable computer. It had just sixty-six electronic parts performing basic logic functions, as opposed to the billions in a modern microprocessor—small enough to easily analyze, but large enough to run a bare-bones neural-network simulation. Using this physical computer, they could compare what IIT says about the simulated network versus the original neural network.

Tellingly, the computer worked in a very different way from a neural network. It was not a highly parallel system in which information was flowing every which way at once. Instead, the computer took one thing at a time. It considered one neuron, then another, then calculated what those neurons would do if they were interacting—but there was never actually any such interaction. Most of the computer's operation was feedforward, implying that it was barely conscious at all. Its modicum of experience was nothing like that of the original network, and it was determined by the hardware, not by the simulation code. "It has nothing to do with what it is simulating," Tononi said. "It could be simulating an avalanche, a hurricane, a brain—it doesn't matter."

Almost all computers today, from your laptop to a supercomputer, are like that, too. They have a degree of parallelism, such as multiple processing cores, but it is nothing compared to that of a neural network. If such a computer were running a full simulation of your brain, it would act like you, but have the experience of being a computer. It wouldn't matter whether the computer was simulating you or a toad; internally it would feel the same. So those who seek digital immortality by cloning their brains in a computer might want to think twice. Conversely, if you have the experience of being a person, you probably are one—you're not trapped in some dystopian science-fiction brain-uploading scenario. Karl Friston, working with the philosopher Wanja Wiese of the University of Mainz in Germany, has explored this issue from the perspective of predictive coding and reached a similar conclusion.[63]

Tononi worries that we face a bait-and-switch situation as computers become ever more lifelike. "The majority of people these days would still say, 'Oh, no, no, it's just a machine,' but they have just the wrong notion of a machine," he said. "They are still stuck with cold things sitting on the table or doing clunky things." But advanced AIs such as ChatGPT and DALL-E are already able to do things that seem to be coming from deeply felt experience, including creating poetry and art.[64] Able to write convincingly on any topic and solve problems in chess, math, and other domains, these systems are starting to demonstrate a generalized intelligence like that of humans. If we judge one of these systems to be conscious, we may accord it the rights of a person. "When that happens—and it shows emotion in a way that makes you cry and quotes poetry and this and that—I think there will be a gigantic switch. Everybody is going to say, 'For God's sake, how can we turn *that* thing off?'"

But, he added, "if IIT is right, that is tragically wrong." Just because machines seem conscious doesn't mean they are. "They may be really imposters," he said. "There is nobody there. . . . You need a theory about what consciousness is to have a proper answer." Late in 2022, David Chalmers reviewed what IIT and other theories have to say about ChatGPT and concluded that it is probably not conscious.[65] The philosopher Susan Schneider of Florida Atlantic University agreed that we need to be careful: "Artificial general intelligence may not be conscious, and that will mean we'll have a case of sapience without sentience." She has speculated that complex, human-level consciousness may be a transitory phenomenon. Having arisen at some point in our evolutionary lineage, it may die out if and when AIs supplant our species: "Consciousness may be a blip, a momentary flowering of experience before the universe reverts to mindlessness."[66]

4

THE QUANTUM BRAIN

IN FALL 1990, WHEN I started my planetary science PhD program at Cornell University, many of my physics professors seemed determined to make science as unfun as they could. Their classes were basically weeders of the sort that are all too common in science and engineering programs at American universities—a form of academic hazing in which professors work you to the bone. Halfway through that first semester, the eminent gravitation theorist and future Nobel laureate Roger Penrose arrived to deliver a series of three lectures based on his new book, *The Emperor's New Mind*.[1] I went at the risk of falling even further behind on my homework.

Penrose offered his own version of what would become known as the hard problem of consciousness. He made the case that a computer algorithm or other mechanistic process can't achieve conscious apprehension: you can follow a procedure to the letter and have no idea what you're doing (an all-too-common syndrome in daily life). So if the brain were governed by the usual mechanical laws of physics, either classical or quantum, it would perform its functions by rote without gaining any insight or self-awareness. He speculated that the new physics required to unify quantum mechanics with Einstein's general theory of relativity might be nonmechanical. If this new physics operated in the brain, it could give our minds their non-mechanical quality.[2] It was invigorating to see a scientist of his stature connect two of the greatest puzzles of modern science: quantum physics and consciousness. And it was great to be among people asking big questions—*this* was why I was in grad school. Later in my studies, whenever I got discouraged, I thought back to those lectures.

I wasn't the only one who felt this way. Penrose's book electrified Stuart Hameroff, an anesthesiologist in Tucson, Arizona. Anesthesiologists turn consciousness off and on for a living, so they tend to have a more than passing interest in what it means. "What Roger said seemed totally bizarre, but also smack on," Hameroff told me. "I just thought, Wow." Together, Hameroff and Penrose—the voluble doctor and the soft-spoken mathematical physicist, an odd couple if ever there was one—hatched a theory for how conscious experience springs from quantum effects in the brain. Penrose laid it out in a follow-up book, *Shadows of the Mind*, published in 1994.[3] The two also launched the first annual conference devoted to consciousness, helping to bootstrap a whole new scientific discipline.

As much as physicists admire Penrose in general, they're not big fans of his work on consciousness. A quip I often heard was that Penrose committed the fallacy of minimizing mysteries: quantum mechanics is a mystery, and consciousness is a mystery, so they must somehow be the same mystery. It was also off-putting that many of the people who flocked to him seemed downright crackpot. So my initial excitement waned, and by the time I got to writing this book twenty-five years later, I had no plans even to mention Penrose. But when I told people I was writing a book about physics and consciousness, they kept asking, "Oh, you mean Penrose?" So I figured I had to say something about his theory. I went to several of the annual conferences that he and Hameroff organize and visited both men at their respective homes. As I delved into their work, my skepticism lessened, and I realized that my earlier dismissal of it had been hasty.

Even if you don't buy their theory, they have done science a service by raising questions that most had swept under the rug. As I mentioned in chapter 1, there really does seem to be some link between quantum physics and consciousness. Our deepest theory of the material world has a place for conscious minds, and by rights it shouldn't. The connection may well be illusory, but even an illusion demands an explanation.

THE MEASUREMENT PROBLEM

As with most great ideas in science, the core principles of quantum mechanics are actually quite simple. It doesn't take mathematics to grasp them, although math does show you the exquisite way these principles mesh together and operate in real-world situations. The theory flows from the observation that material objects behave in two apparently contradictory ways.

On the one hand, matter and energy are particulate. They come in chunks, which is the meaning of the word "quantum": a discrete unit. In the case of light, these chunks are photons—particles of light. If light is dim enough, it appears as individual flashes, which our eyes are potentially sensitive enough to see.[4] On the other hand, matter and energy are also wavelike. They spread out. They surge through gaps, part around obstacles, and bend around corners. They can be added or subtracted to form new waves.

Neither particulate nor wavy behavior is strange on its own. It's their coexistence that makes physicists' heads explode. Wave behavior seems to be the more fundamental of the two, at least in the standard picture of quantum mechanics created by Erwin Schrödinger. Most of the time, matter and energy are wavelike. Whenever we observe them, though, they are particulate. For instance, an object can exist in a wavelike state, of being spread over a wide area, but on measuring its position, we find it in only one specific place. The underlying wave behavior, which physicists represent mathematically as a "wave function," is not directly visible to us; it must be inferred from multiple observations and, depending on the size and type of object, can be very hard to discern. Quantum objects are like kids on a playground. They get up to all sorts of mischief, but make sure to be on their best behavior whenever a parent or teacher is watching.

Textbook quantum theory, first articulated by John von Neumann in 1932, describes this two-faced behavior using two laws: a special rule that applies at the time of measurement, and a wave equation that governs the object the rest of the time.[5] According to the latter, the wave function associated with the object, though not quite the same as an ordinary wave, behaves in many respects like one. It spreads out and moves around in a continuous motion. But according to the former, when you measure the object's position, the wave function instantly collapses to a single location. The wave function describes properties other than position, too, notably momentum and spin. The object can juggle multiple options for any or all of these properties—this is the state of superposition that I mentioned in previous chapters—until it is measured, whereupon it collapses to just one, like choosing from a menu. The choice is random. Each possible outcome can occur with a probability given by the height of the wave function at that location. In combining smooth propagation and abrupt collapse, quantum mechanics is a meeting of opposites. "It's like a marriage—a perfect marriage," Penrose quipped to me.

Wave function collapse is weird, bordering on paradoxical. It is irreversible, the only process known to physics that can't be undone even in principle. It is also unpredictable; until a collapse, quantum physics is completely deterministic—the state of affairs at one time fixes the state at all other times, leaving no room for chance—whereas collapse spins the wheel. And it is, for want of another word, magical: if you so much as look at something, you collapse its wave function—but there is no underlying causal mechanism. It just happens.

Magic, per se, doesn't bother physicists. They simply roll it into their theories and start calling it physics. That's what they did with gravity and magnetism, both of which were once thought to be supernatural.[6] The real problem is that the collapse rule is intimidatingly vague. To apply it, you have to know what a measurement is, and although that may seem obvious at first, it quickly gets murky. Does any interaction between two systems qualify? If someone else makes the measurement and doesn't tell you the result, is that a measurement for you, too? How do you avoid circularities, such as measurers who measure themselves?

For example, imagine you shine a flashlight at a half-silvered mirror, like the one-way mirror in a police interrogation room. Half the light goes through the mirror and continues on its way, while the other half is reflected back or off to the side. This splitting is classic wave behavior, also seen in water waves, sound waves, seismic waves, and so on. To study this at the particle level, you dim the flashlight enough so that it emits only a trickle of photons. A particle, unlike a wave, does not split under these conditions. When you place sensitive detectors on either side of the mirror and watch their electrical outputs, you will find that half the photons are reflected and half are transmitted. Each photon chooses a path at random.

But at what point does a photon make its choice? Not when it strikes the mirror: in the quantum world, each photon both passes through and reflects off it at once. Not when the photon is picked up by a detector: each detector both registers a photon and doesn't. Not when electrical signals from the detectors are recorded by a computer: the computer stores both a 0 and a 1 in its memory. Not when the computer displays a value, not when light from the computer screen enters your eye, not when nerve signals travel from your retina to your visual cortex. At no point in this long chain of events is there any collapse. Each link enters and remains in a superposition, encompassing multiple possibilities. All these events are described by the wave equation, and there is never any cause to employ the collapse rule.

In the movie *Sliding Doors*, Gwyneth Paltrow's character either makes her train or doesn't, which causes her either to break up with her boyfriend or to stay with him, either to flourish in her career or flounder, and either to get hit by a white van or not. The initial ambiguity infects everything that follows. So it is with quantum measurement. The particle's ambiguity—its superposition—expands to include the detector, computer, eye, and brain, binding them together in collective indecision, a bond that physicists call quantum entanglement. "It seems as though nothing can ever be settled by such a measurement," the physicist Hugh Everett wrote in 1956.[7]

There is only one place where physicists know that measurement yields a definitive result: in the subjective experience of the observer. By its very nature, our experience is unified. We do not see mutually

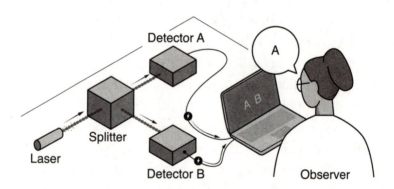

4.1. VON NEUMANN CHAIN OF MEASUREMENT. In 1932 the mathematician John von Neumann conceived of quantum measurements as a sequence of operations. In one of the simplest examples, you fire a photon at a beam splitter, such as a half-silvered mirror. It could pass straight through or be deflected off to the side. Detectors will register which path it took and relay the information to a computer, which displays the result. You have a fifty-fifty chance of seeing it take one path or the other. Now here is the funny thing. Quantum theory says the photon took *both* paths, both detectors fired, and the computer displayed both outcomes—an ambiguous state known as superposition. Yet you see only one outcome. Why is that? Subjective experience is the only thing known to science that does not enter superposition.

conflicting eventualities. We do not see a photon both pass through and reflect off a mirror—we see it do one or the other. "The being with a consciousness must have a different role in quantum mechanics than the inanimate measuring device," wrote the physicist and future Nobel laureate Eugene Wigner in 1962.[8] Subjective experience thus stands outside the quantum order as the only known phenomenon in nature that is never superposed. The waves that ripple through the quantum ocean break on the shores of our conscious selves.

SO THIS IS how textbook quantum physics gives the mind an essential role. Very few physicists are happy with that. How can a theory with pretensions to being fundamental require sentient observers? Such a theory is supposed to explain our existence, not presuppose it. We don't even really know what consciousness is; writing it into the laws of physics is like building a house on quicksand.

Physicists console themselves that they can still apply quantum theory to most practical situations. They can say that collapse occurs somewhere—it doesn't really matter where, as long as it doesn't affect the phenomenon they're studying in that experiment. For instance, in the case of the half-silvered mirror, they can suppose that the detector causes the photon to collapse. Technically, that's not true; in fact, the detector itself will go into a superposition. But quantum–optics researchers usually aren't studying the detector itself; it's the photon they care about. So they can get away with drawing an arbitrary line, known as the Heisenberg cut, between object (the particle) and subject (the apparatus and human observer). This stratagem doesn't always work, however. In cosmology, for instance, the object of study is the whole universe. You can't draw a line between the universe and yourself, because you're part of it. Line drawing also gets tricky when two observers are observing each other: each person puts the other on the wrong side of the line, creating a contradiction.

If you ask physicists and philosophers about this, many will say something like "There's no problem, don't be silly. The answer is . . ." and then offer a solution that sounds entirely plausible—until, that is, you talk to the next person, who offers a completely different reply, delivered with equal self-assurance. That the debate remains unsettled after nearly one hundred years tells us the answer can't be easy. One lesson about quantum mechanics is to be wary of anyone who is a little too sure of themselves. Let's consider several common responses to the measurement problem and how they fall short.

QUANTUM DOESN'T MEAN SMALL

To begin with, could collapse be a question of size? Might objects behave unambiguously as long as they are big enough? This is a powerful intuition for many physicists. Alas, it's not what quantum theory says. Because Einstein, Schrödinger, Niels Bohr, and their contemporaries developed the theory to solve microscopic mysteries such as why atoms don't implode, it became typecast as a theory of little things. Even today, that's how some physicists and science writers portray it. Yet its pioneers quickly realized that the theory has no size limit. Quantum effects are easier to conceptualize in the case of individual particles, which are easier to control experimentally. But the theory applies equally to masses of particles. Within quantum theory, nothing is immune to superposition, no matter how big or complicated. That includes humans and other sentient beings. We don't see human beings in some weird superposition state, of course, and the question we're trying to answer is why not.

For size to be the deciding factor, quantum theory itself would have to fail at some level. And maybe it does. Penrose, for one, thinks so.[9] I will delve into his views later. But so far there is no evidence that quantum mechanics ever fails. Experimenters have checked for exceptions and size thresholds.[10] Using variants on the half-silvered mirror experiment, they have placed large molecules,[11] tiny tuning forks,[12] photosynthesizing bacteria,[13] and dormant tardigrades[14] into superpositions. These weren't superpositions of transmission through or reflection by a mirror, but of some other property, such as the distribution of electric charge in the tardigrades' little bodies. Furthermore, experimenters have created distinctively quantum forms of matter, such as superconductors, that are big enough to hold in your hand. Quantum mechanics has no size limit as far as they can tell.

IS SUPERPOSITION ALL IN YOUR HEAD?

A second response is that superposition is an artifact of ignorance. With the half-silvered mirror, you might suppose that each photon either passes through or is reflected back, never both, and the twist is that you don't know which, so you have to consider both options until you do. The supposed collapse is all in your head, like thinking you got both a quadcopter and a PlayStation for your birthday until you tear open the wrapping and see the drone.

Such an explanation is dicey. Measuring a particle is not like unwrapping any gift you've ever seen. To make the gift metaphor work, you'd have to imagine that the gift giver had no clue what's in the box, either. You'd also have to imagine that two gift givers could somehow link their gifts so that, despite not knowing what's inside, they avoid giving you the same thing. These are some of the shenanigans that quantum particles are capable of. They seem to require that the outcome be genuinely open until the final reveal. We know this because the physicist John Bell showed in the '60s that particles can exhibit patterns that would be hard to explain if their properties were fixed in advance.[15] By now physicists have dozen of examples of such phenomena. Because the patterns are typically statistical, physicists like to compare them to coin tosses and other games of chance. Sometimes these scenarios seem contrived, but it's all in the service of nailing down the mathematical proof.

In one especially elegant experiment that the Nobel laureate Anton Zeilinger of the University of Vienna told me about, he and his colleagues made particles run a kind of obstacle course. They prepared a photon in a special state and aimed it at an assortment of polarizing filters to see whether it would pass through them.[16] They repeated the experiment multiple times. On each run, they chose two of five available filters and recorded whether the photon made it through. Then they chose another two, and another two. This experiment was mathematically equivalent to a type of shell game: Arrange five cups in a circle, hide coins under some of them, and ask a friend to pick up any two adjacent cups. Suppose you place coins under the first, third, and fifth cups. Your friend picks up the first and second, and sees a single coin. Then she picks up the second and third: again, a single coin. She continues in this vein. But when she gets to the fifth and first cups, closing the circle, she should see two coins. The pattern breaks—it has to when you have an odd number of cups. Usually we express this in terms of probability. Your friend should see a single coin 80 percent of the time; these odds directly reflect the fact that, before your friend picked up any cups, the coins were already in place.

But when Zeilinger and his team played the quantum equivalent of this game, they saw a single coin about 90 percent of the time. That's astronomically unlikely if the outcomes were determined in advance. Ergo, they were not determined in advance, but on the fly. Superpositions persist until the moment a measurement is made.

In keeping with the lesson about never being too sure of ourselves, I should acknowledge that not everyone accepts this conclusion—there are ways to think of superpositions as illusory. But they involve their own weirdnesses and are definitely not an easy out. The leading such approach, developed by David Bohm, involves instantaneous action at a distance.[17] It may well be the way to go. My point is simply that you can't say superposition is "just" our ignorance of the true state of affairs.

DECOHERENCE

A third response to the measurement problem is that the second quantum law implies the first: that the evolution of waves can naturally produce a collapse of sorts. This ersatz collapse goes under the rubric of what physicists call decoherence. The basic idea is that a quantum wave is never isolated from its surroundings. As it propagates, it encounters other matter and becomes so jumbled that it is no longer perceptible as organized wave motion. Thus a particle appears to lose its wavelike qualities, including superposition, as if it had collapsed.

Through decoherence, a localized infection of ambiguity snowballs into a pandemic. The particle's superposition spreads not only to the measuring apparatus and your brain, but also to your entire body, to the air in the room, to the building, and ultimately to the entire universe. This unstoppable spread of quantum superposition can look, to someone caught up in it, like collapse. For instance, it is all but irreversible. By the time the infection reaches global proportions, there's no undoing it.

Decoherence also addresses an important and often unappreciated aspect of quantum measurement, which you might call the menu problem. When they collapse, particles do not choose their final state from among limitless options; their options are highly structured. Yet quantum mechanics itself provides no menu. In the example of the half-silvered mirror, I've been assuming the two options are to reflect off the mirror or to pass through the glass; by the end of the experiment, the photon is on one side or the other. But that's not a given. The theory works equally well with less intuitive options such as "a little bit reflected plus a little bit transmitted" and "a little bit reflected minus a little bit transmitted," in which the photon winds up straddling multiple locations. Because quantum theory doesn't specify the menu of options, that task must fall to some other physical process, and decoherence performs this function. Because decoherence typically involves particles making direct contact, it defines a menu of distinct spatial positions, which matches our classical intuitions.

But decoherence doesn't touch the central puzzle, and its originators never claimed it did.[18] The outcome of a measurement is still a multiplicity rather than one particular answer, leaving physicists at a loss to explain why we see a single result. Furthermore, decoherence requires that you differentiate "object," "observer," and "surroundings," which reinserts the observer into the picture.[19] "Decoherence is dependent on the perspective of the self onto the world," said Heinrich Päs, a theoretical physicist at the Technical University of Dortmund.

However much physicists might wish to eliminate the mind from their basic theories, it's not clear that they can. "The core issue is that we really don't understand conscious experience," said Mile Gu, a theoretical physicist at Nanyang Technological University in Singapore. "We don't really understand what it means to experience something. And when we don't understand that, it's very hard to make any concrete statement about a physical theory that involves conscious measurement." Whether or not observers play some direct physical role, interpreting quantum theory at the very least requires thinking about how our minds perceive the world and reason about it. "Consciousness is fairly deeply ingrained in a lot of these interpretations," said Gu.

DOES THE MIND CAUSE COLLAPSE?

Though widely considered to be a fringe notion, the most straightforward conclusion is that the mind really does cause collapse. Even Wigner, who spoke up for the idea in the '60s, later backed away from it.[20] But it hasn't always been deemed nutty. In 1939 the theoretical physicists Fritz London and Edmond Bauer, who worked at nearby institutes in Paris, outlined a role for consciousness.[21] They didn't feel they were going out on a limb; they thought they were just filling in the orthodox interpretation of the theory.[22] London had made his name creating the first quantum theory of supercon-ductors, proving that quantum effects show up at large scales, so it stood to reason that humans were fully part of the quantum world and thus of any experiment. The outbreak of war disrupted the two men's work: Not long after finishing up their paper, London paid a bribe to get on a ship bound for New York. Bauer stayed on in France, and his four children fought in the Resistance.[23]

As with many other ideas in physics that fall out of favor, the problem wasn't plausibility so much as vagueness. If you don't know what consciousness is, how can you build a theory of physics on it? That's where emerging theories of consciousness can now help. They make concrete predictions about when a lump of matter is conscious.

In 2013 David Chalmers and fellow philosopher Kelvin Mc-Queen, both then at the Australian National University, began to use integrated information theory (IIT from chapter 3) to clarify the collapse rule. They trailed the idea for years in talks both to physi-cists and to neuroscientists, which is how I heard of it, and finally published a paper in 2022.[24] Several other physicists and philoso-phers also took up the idea in the meantime and offered their own analyses.[25]

Their proposition is that a conscious system stands outside quan-tum physics. If you try to put it into a superposition, it will fight back. "It will resist this superposition," explained McQueen, who is now a professor at Chapman University in Orange, California. In the taxonomy of solutions to the measurement problem, this conjec-ture falls into the category of supposing that quantum theory fails at some level. The idea is that highly interconnected systems such as our brains violate it.

Chalmers and McQueen and the others formulated their theo-ries by repurposing the equations that physicists over the years had proposed for setting a size threshold for quantum mechanics. They replaced that threshold with the IIT general measure of consciousness, Φ, as well as its criteria for specific conscious experiences. They calibrated the equations so that a simple neural network, with zero or low Φ, obeys quantum mechanics to the letter, whereas an intricate network deviates. "Systems with sufficient integrated information do not respond to entanglement by superposing, but by collapsing the entangling system," McQueen said.

The way it works is that, during a measurement, the experimental apparatus connects a particle or other object to the brain. We normally think of measuring as transferring information from a particle to our brain, but in these theories the traffic is two-way. Through the connection that the measurement system establishes, the mind reaches out, grabs particles that are poised between possibilities, and tells them, Choose!

One reason it took Chalmers and McQueen so long to publish their paper is that they had to work out the menu of options that particles choose among when they collapse. Because they are postulating a real collapse rather than an ersatz one, they can't assume that the process of decoherence sets the menu. Instead, the menu must be determined by the conscious experiences that our brains are capable of. The particle will settle into a state that is consistent with a specific pattern of neural connectivity. McQueen and Chalmers don't explain why that would translate into seeing the photon either here or there—why, in other words, our experience is of localized objects. Presumably this has some practical rationale. Our conscious experience is shaped by evolution and by education to represent the world in a way that helps us to survive, which entails seeing the world as laid out spatially. The cosmologist Max Tegmark has also suggested that IIT itself could provide an answer to the menu problem— the world may appear to be made of separate but interacting parts because our own minds, according to IIT, are made that way.[26]

This raises the mind-blowing possibility that other sentient beings might think in vastly different categories. Päs calls them "quantum aliens."[27] Depending on how their minds are structured, they might perceive the fundamental unit of space to be a line rather than a point, so to them, there would be nothing weird about existing at multiple locations at once. Instead, they might find it mystifying that anything could ever confine itself to just one spot. The science-fiction writer Ted Chiang, in "Story of Your Life," imagined extraterrestrial visitors who see all of time laid out before them. Per the IIT-based theories, these aliens would have a different effect on the particles they observe than we do.

Another fascinating issue that McQueen and Chalmers comment on is that collapse takes time, so there is a brief window during which the brain is in a superposition of two conscious experiences. What would that even mean? They explore several options. Perhaps it would be a kind of split-brain syndrome in which two independent minds occupy the same brain—an option that physicists have considered in the past and that I will return to in chapter 5. But they lean toward thinking of it as a bizarre new category of experience, like synesthesia or an LSD trip, that merges qualities of more familiar ones.

To test the IIT-based theories, McQueen and others have proposed adapting existing quantum threshold experiments so that experimenters can compare objects with different degrees of integration. Even a tiny quantum computer might do—small though it is, it might still have a high enough value of Φ to quickly collapse any particle it came into contact with. "What's nice about that is that the theory can be tested without needing to test on large quantum systems," McQueen said. Physicists who do threshold experiments are receptive: "We plan to go along this direction of study as well," the quantum researcher Cătălina Curceanu at the Istituto Nazionale di Fisica Nucleare in Rome told me in 2021.

At the end of the day, you might still wonder how and why the mind would exert a collapse effect. McQueen said he sees his and

Chalmers's theory as merely a stepping-stone—first, clarify the circumstances of collapse; later, worry about the underlying mechanism. Perhaps the collapse is triggered not by consciousness or information integration per se, but by deeper physics that integrated systems are somehow more sensitive to. "The immediate goal is self-consistent description," McQueen said. "But the process of reaching such a goal can often lead unexpectedly to new kinds of explanation."

IMMUNE TO SUPERPOSITION

Penrose comes at the problem of quantum measurement from the opposite direction. Instead of supposing that consciousness causes collapse, he argues that collapse causes consciousness. He starts by identifying an entirely objective mechanism for collapse—"It takes place in the physics, and it's not because somebody comes and looks at it," he told me—and then he mulls what such a mechanism might mean for our mental experience.

The mechanism he proposes is gravitational. In our current understanding, gravity is produced by a field akin to the electric or magnetic field. This field is a structure that pervades all of space (and indeed is an aspect of space). When we say that Earth exerts a force on an apple, what we mean is that the planet affects the field, which in turn acts on the apple. Most physicists think that the gravitational field is as quantum as anything else in nature—this has been their starting assumption in seeking a unified theory of physics. Penrose thinks the unification project has stalled because it presupposes that quantum physics is the more fundamental concept and that gravity must somehow fit into its framework, whereas he thinks that gravity might alter the quantum framework.[28]

Building on work in the '60s by the physicists Richard Feynman and Frigyes Károlyházy, Penrose suggested in the '80s that the gravitational field stands outside quantum physics, unable to remain in a superposition for very long, or at all.[29] Whereas an ordinary particle will remain in superposition forever if it is kept isolated, the gravitational field will quickly collapse. That in turn affects planets, apples, and anything else the field touches.

Penrose's theory adds a new element to the textbook account of measurement. During a measurement, the particle's superposition spreads to the measuring equipment, eyes, and brain. If the equipment has a dial, the dial might point to multiple positions at once. Within the eyes and brain, ions and proteins might move to both the left and the right sides of a cell. Thus the widening superposition will affect the arrangement of mass in the laboratory. Because the force of gravity is a function of mass, this superposition threatens to put the gravitational field into superposition, too. On Penrose's account, that's where superposition must stop. Being immune to superposition, the gravitational field settles into one state or the other, and once it does, it resolves the ambiguity of everything else. The brain, eye, equipment, and particle settle down, too. "Gravity does not like superposition," Curceanu explained. "Gravity is classical! So it reacts back on the wave function."

So far, this approach is actually quite similar to Chalmers and McQueen's. Both speculate that the process of observation causes a particle to collapse, not because of some mystical power that the mind has, but because connecting the particle to a larger system exposes it to some new kind of physics, be it information integration or gravitational effects. Where the approaches diverge is in the neuroscience. While Chalmers and McQueen adopt IIT, Penrose developed an entirely new theory of consciousness. Having proposed new gravitational physics to explain quantum collapse, he speculated that this new physics might have something to do with the mind. His reasoning was straightforward: if the known laws of physics can't explain our minds, you apparently need new laws—and hey, here is a new law. As a physicist, however, Penrose didn't have much to add about the neuroscientific details. That's where Stuart Hameroff entered the story.

While a medical student in the early '70s, Hameroff spent a summer in a lab doing research on cancer biology. One thing that can go awry in the cells of cancer patients is mitosis, the process of cell division. During mitosis, a cell makes copies of its chromosomes, pulls the copies apart, and splits in two. Hameroff became fascinated by what does the pulling. This was his introduction to microtubules.

These miniature filaments had been discovered in the '50s.[30] Composed of a protein called, appropriately, tubulin, they are the bones of a cell's miniature skeleton. Watching them choreograph cell division, Hameroff thought that they seemed awfully smart for bones. Indeed, biologists at the time were finding that microtubules perform a range of surprisingly sophisticated functions. Cilia and flagella, the little tentacles of a cell, are made of microtubules. Microtubules give the cell a primitive sense of touch and smell, let it differentiate among stimuli, and fuse information from multiple sources.[31] Tubulin molecules can change in shape, and electric charges moving back and forth within them can store information.

Hameroff contributed to the study of microtubules in the '80s by working out how these structures have all the makings of tiny computers.[32] Most neuroscientists, he concluded, had been looking at the wrong level of the brain to understand how we think and feel.[33] "Neuroscience tells us that neurons are dumb—nodes in a network—and you get consciousness from networks," Hameroff said. He came to think that the basic unit is not the neuron, but the microtubule, that the arrangement of microtubules within neurons may be as or more important than the connections among neurons. This was a controversial claim—arguably even more so than Penrose's ideas about quantum effects. In recent years, neuroscientists have come to agree that neurons are complex computers in their own right, containing what amounts to a miniature neural network.[34] Whether microtubules are involved, though, remains contentious.

For microtubules to serve a computing function, they'd need, as all digital computers do, a clock to synchronize their signals. Hameroff turned to a mechanism proposed by the physicist Herbert Fröhlich in the '60s. Fröhlich argued (and experimenters later proved) that, using microwaves or mechanical vibrations, you can cause protein molecules to oscillate in a coordinated way, overcoming the chaotic motions that ordinarily prevail inside a cell.[35] Hameroff thought such oscillations could serve as a computer clock—a bit like a quartz clock, in fact. He suggested that general anesthetics knock you out by stopping the clock. That would mean the oscillations help to make us conscious.[36] Fröhlich's comparison of the oscillations to a quantum process known as Bose-Einstein condensation gave Hameroff his first inkling that quantum effects might be important to consciousness.[37]

When he read Penrose's book in 1992, Hameroff decided he was holding the missing puzzle piece. Quantum effects, including the ones Penrose was proposing, tend to be most prominent at small scales. Microtubules just so happen to be the right size. Hameroff reached out to Penrose, and the two hit it off. They were part of a minitrend. Around the same time, other scientists—most prominently John Eccles, who had won the Nobel Prize for studying the transmission of neural signals across synapses—were also suggesting that quantum effects might create our conscious experience.[38]

QUANTUM PANPSYCHISM

What Penrose and Hameroff proposed is that each of our conscious experiences is a quantum collapse. Our minds don't cause collapse, but are constituted by it. When we see red or hear a minor chord, quantum wave functions are collapsing to give these experiences a quality beyond their bare information content. Why that should be the case, Penrose and Hameroff do not claim to be able to explain. Rather, they take it as a primitive feature of the universe. "Consciousness in some way has been there all along," Hameroff said. They are proposing, in short, a form of panpsychism, the old doctrine that mentality is ubiquitous in the natural world.[39]

But it's not just any form of panpsychism. It's an original version that evades some of the objections to panpsychism that people have raised over the decades. For instance, detractors argue that fundamental physics has already enumerated all the primitive features of the universe, leaving no room for a conscious ingredient. In other words, they claim that physics is "causally closed," meaning that nothing lies outside its capacity to explain. There's a slight problem with this critique: physics *isn't* causally closed. At least one thing still lies beyond the reach of our theories: quantum collapse. And that one thing is peculiarly connected with consciousness. So Penrose and Hameroff have slotted panpsychism into the one place in modern physics where it could clearly fit.[40]

By invoking quantum effects, Penrose and Hameroff also neatly fix the main flaw of panpsychism, the combination problem, which also came up in chapter 3 where I talked about IIT. How do primitive conscious experiences meld into a single human mind? Why are we not trillions of minds occupying the same body? Quantum mechanics suggests a couple of ways to create one mind out of many. Penrose and Hameroff focus on quantum entanglement. The standard example of this phenomenon is a pair of photons. In laboratory experiments, such a pair is typically created when an atom gains energy and releases it, or when a laser beam amplifies random fluctuations in an electromagnetic field. By virtue of their common origin, the two photons bear a fixed relation to each other. They might have the same color, or the same polarization, or opposite polarizations—it depends on the setup, but in every case, if you know what one photon is like, you know what the other is like, too.

By taking advantage of the inherent uncertainty of quantum physics, experimenters can deliberately build some ambiguity into the creation process, and the photons inherit it. They emerge in a superposition of two or more possibilities rather than with well-defined properties of color and polarization, for example. When you measure either of the photons, you will see it settle into one of these possibilities at random. But although the photons are individually ambiguous, the pair is collectively unambiguous. Each will settle into a color or polarization at random, but it will be the same random for both. They synchronize themselves without any communication or mechanism.

How they do that is one of the deepest puzzles in science, but for now we can set that aside and focus on what entanglement might mean for consciousness. In effect, entanglement takes two or more particles and makes a single larger structure out of them. Importantly, this structure has properties that cannot be reduced to those of its parts. The whole is not just more than the sum of its parts, but something entirely different. Strictly speaking, there aren't "parts" anymore. And if you can do that with conscious experiences, you will ensure that a single mind results, thus solving the combination problem.

A third distinctive feature of Penrose and Hameroff's version of panpsychism is that they don't imagine that trees or rocks or electrons are conscious. They adopt an approach that philosophers refer to as panprotopsychism.[41] Quantum collapse in their view provides an ingredient for consciousness, one that is otherwise missing from physics, but not everything will avail itself of it. Rocks, electrons, and, probably, trees don't have the right neurological inner workings.

In particular, Penrose and Hameroff distinguish experience from the experiencer. A quantum collapse, also known as state reduction, is a fleeting, unarticulated experience, not necessarily felt by anyone or anything. "These protoconscious events due to objective reduction—happening in the environment willy-nilly, all the time, just occurring here, there, and everywhere—would be random and isolated," Hameroff explained to me. "They wouldn't be entangled or connected with any other, so we call those 'protoconscious.' They're happening ubiquitously all the time. They come and they go. Roger [Penrose] suggested that they would have some quality of experience, or qualia. Each one would be some quale of experience, but it would be random. It might be a good feeling, a bad feeling, some type of experience."

On this view, consciousness is a constellation of discrete, spontaneous events that go off like camera flashes in a dark auditorium. The auditorium does not produce them; it is just the place where they happen to occur. Penrose and Hameroff do not think the brain, or anything else, generates conscious experiences. It is a place where they occur. Another metaphor might be a garden. The gardener plants the seeds, lays the mulch, and weeds out the crabgrass. But the rest is up to the begonias. Your brain is the gardener. It creates the right conditions for and orchestrates the otherwise disconnected flashes of experience to form a stream of consciousness and a narrative self. According to Penrose and Hameroff, the brain creates these conditions in three steps.[42]

First, there is a preparatory phase. The microtubule structure inside a neuron acts as a computer, assimilating information from the senses and memory. All the tubulin molecules become mutually entangled. That way, when they collapse, they collapse as one. By virtue of its size and mass, this cluster of molecules collapses faster than the molecules on their own would. "I think of it as a quantum computation that halts or terminates, not by someone else making a measurement, but by self-collapse," Hameroff told me. Penrose and Hameroff attribute the collapse to gravity, but their theory of consciousness doesn't really depend on the specific trigger. Physicists have proposed others and they'd work, too.[43]

Second, the collapse occurs and, with it, a flash of subjective experience. This could happen as often as 10 million times per second, Penrose and Hameroff estimate. The collapse chooses one option at random from the menu established by the preparatory phase; that menu determines what the experience is an experience of, such as color or pain. "The content comes from the pattern that's occurring in the computation and the orchestration," Hameroff said. He compared it to "a palette of dabs of paint, and the artist—or, in this case, the microtubule orchestration—selects a dab of this and a dab of that."

Third, as a result of the collapse, the neuron either will or won't fire, or a synapse might become either more plastic or more rigid. In this way, conscious experience is part of the functioning of the brain. That said, the brain does not really need consciousness. Indeed, most of our brain activity is unconscious. Hameroff argued that the brain evolved to provoke and string together moments of consciousness, not because it helps us process information, but because consciousness gives us a sense of self-worth.[44] "The organism wants to survive," he said.

DISRUPTING THE THEORY

When I talk to physicists about Penrose and Hameroff's theory, their most common objection is that neurons are hostile environments for quantum effects. Max Tegmark—the cosmologist who is otherwise a fan of physics-based theories of consciousness—published a widely cited takedown in 2000.[45] He argued that the process of decoherence inside a living cell would wreck the preparatory phase. As swarming water molecules and ions bombard the tubulin molecules or exert electrical forces on them, the carefully constructed entangled superposition would not last even as long as a trillionth of a second. So even if gravity caused quantum collapses, and those collapses constituted conscious experiences, the brain could never control them, and we'd have no coherent stream of consciousness.

Tegmark's paper came like a bolt from the blue. Hameroff said he and Penrose were visiting Washington, DC, at the time it was published. He recalled: "When Tegmark's paper came out, I get an email from [a colleague], 'Well, that settles that.' . . . People I knew, people at the university, said that was devastating to our theory. Here's Max Tegmark . . . he was the boy wonder of physics." Penrose printed out the paper to read that night, and they talked about it over breakfast the next day.

They realized that the critique wasn't all that new. Penrose himself had acknowledged the potential for disruption in his 1989 and 1994 books.[46] In response, the two men and their coauthors developed the idea that some types of superpositions can tough it out.[47] For instance, the oily regions of a cell are isolated from the watery solution and therefore protected, somewhat, from molecular bombardment.[48] And quantum information would be even better protected if it could worm its way into atomic nuclei.[49] In the years since Tegmark's analysis, experimenters building quantum computers have come up with all sorts of ways to shield superpositions; evolution may have stumbled upon those methods, too. For instance, physicists and biologists have made a good (if by no means airtight) case that quantum superpositions can persist in photosynthetic molecules and perhaps help them to squeeze all the available energy out of sunlight.[50] So Tegmark's critique has lost some of its bite.

"Everyone assumed that quantum biology is impossible," Hameroff said. "The brain is too warm, wet, and noisy. . . . From '94 to 2006, quantum biology was considered crazy. So in 2006, lo and behold, [researchers found that] photosynthesis at ambient temperatures utilized quantum coherence, and if not for the efficiency it gives us, we might not be here; there might not be enough food. So it's essential for life. If a potato or rutabaga can figure out how to use quantum biology, maybe our brains can figure it out."

So far the evidence for this is scant. Penrose and Hameroff's theory has so many novelties—quantum collapse due to gravity, processing of data inside neurons, unconventional panpsychism, and shielding from molecular bombardment—that it has to be taken as a dark-horse candidate for an account of consciousness. Still, this isn't one of those philosophical arguments that will never be resolved. It is well within the ability of experiments to confirm or refute. Scientists can check whether microtubules really do have anything to do with brain function and whether quantum physics breaks down in certain settings, as Penrose and Hameroff postulate.

Their theory, like predictive coding and IIT, doesn't have to be swallowed whole. Even if quantum effects have nothing to do with consciousness, they might perform other functions in the nervous system. Some quantum phenomena might actually thrive in the chaotic innards of cells.[51] Mile Gu has studied a class of such phenomena known as quantum discord. "There are potential means for quantum effects to be useful in the brain even if we don't have full entanglement," he said. "But it's a wildly open question whether the brain actually uses those effects at all."

In many situations, the disruption of quantum effects by the environment offers a positive benefit: it forms the basis for exquisitely precise sensors. It could explain birds' biological compass, for instance.[52] When a bird's retina absorbs light, chemical bonds break and a pair of electrons is torn asunder. Like photons in the experiment I described earlier, these electrons are entangled by virtue of their common origin. They eventually reunite, but in the meantime, they are affected by magnetic fields. The electrons' magnetic response modifies the chemical reactions they are later involved in, and the varying results of these reactions correspond to some orientation vis-à-vis Earth's magnetic poles. A similar magnetic process may operate in the human body, and even if it doesn't give us the ability to tell where north or south is, it might make our cellular processes sensitive to internal magnetic fields, perhaps altering how drugs act in us.[53] In a different way, our sense of smell may be quantum, too.[54]

Engineers might also be able to exploit these quantum effects in AI systems. In 2006 Elizabeth Behrman and Jim Steck, the quantum-neural-network pioneers I mentioned in chapter 2, and their colleagues showed that engineers could build a quantum Hopfield network out of microtubules.[55] They'd have to keep it cold, approaching absolute zero, which is entirely doable in a machine system. Behrman told me: "We absolutely might be able to do bioengineering to make something like the microtubulin net."

BUILDING A COMMUNITY

Penrose and Hameroff have, if nothing else, helped to legitimize the scientific study of consciousness. Their conferences embrace a tolerant spirit that seems appropriate to a nascent discipline. Academic conferences in general can be lots of fun—astronomers, in particular, have an underappreciated talent for dancing at afterparties. But Penrose and Hameroff's event is in a league of its own. In the exhibit hall, you can try on EEG skullcaps or advanced virtual reality headsets; kick off your shoes to climb into a flotation tank; do yoga by black light; or be coached in meditation by an android. The main lectures combine physics, brain science, medicine, AI, and philosophy; few other scientific events range across such a broad range of disciplines. Prior to the pandemic, Penrose—now hard of hearing and reliant on opera glasses to read lecture slides—still attended, giving pretty much the same talk every time but strolling around to hear from young people and encourage their explorations. It's not often you see a Nobel laureate mingle with the public so unpretentiously. After hours, there are, I am told, experiments with consciousness of an unofficial kind.

At one of these conferences, I had an hours-long dinner with Mani Bhaumik, an Indian American physicist who is a force of nature in his own right. Born into a low-caste Bengali family, he helped to invent the excimer lasers used in eye-correction surgery. Then he moved to Hollywood, hosted raucous parties, and, for a while, dated Eva Gabor. Eventually he cycled back to physics and founded an institute at UCLA. Bhaumik is grateful to Penrose for setting an example of intellectual openness. "Through his two seminal books on consciousness, he managed to bring the skeleton of consciousness out of the physicists' closet," he said.

"Coming here, in this context where everybody agrees at least that we have a problem, it makes you feel better," agreed Marina Vegué Llorente, a mathematician who took up computational neuroscience and is now doing a postdoc at Laval University in Quebec. "You say, I'm not the only crazy person in this hall. It's true that when you are in other contexts, more typical neuroscientific contexts, many times you have just to shut down, because otherwise they say that you are not scientific enough."

5

PHYSICS IN THE FIRST PERSON

IN CHAPTER 4, I BEGAN to explore the puzzle of quantum collapse and the peculiar role that our conscious minds appear to play in shaping reality. Roger Penrose and others think the answer is to patch up quantum theory, but as far as experimenters can tell, quantum mechanics brooks no exception. So a more prudent option would seem to be to take quantum physics as it is rather than seek to alter it. When you see where this path leads, though, you discover it is just as radical as anything Penrose has proposed. In particular, it raises unsettling questions about the objectivity of knowledge.

The trouble is most acute in a special type of meta-experiment in which observers observe other observers. If you set up the experiment right, quantum theory makes conflicting predictions about the outcome. "Paradoxical situations arise when the measurements are performed on the agents themselves," said the quantum physicist Časlav Brukner of the University of Vienna, a leader in this line of investigation.

Observers in these experiments are working from the same theory, do everything perfectly, and still come to blows. To many physicists and philosophers, that suggests that quantum mechanics is inherently observer-dependent, that the equations give us, at best, a first-person perspective. Some think a third-person perspective is buried in quantum theory, if only we can dig it out; some think we need a new theory altogether; and some think no third-person perspective is possible, period. Physics seems like one of the last redoubts of absolute fact, yet quantum mechanics may take even that away from us.

THE META-EXPERIMENT IS similar to Schrödinger's cat, except that it focuses on what it would be like to be that cat. After talking it up for years, Eugene Wigner published a paper on this scenario in 1962; by then, Hugh Everett, who took his classes as a grad student, had already built his dissertation on it.[1] Officially the setup is known as the Wigner's-friend thought experiment, and as with a lot of quantum physics experiments, it is simple yet profound.

Consider again the experiment in which you shine a flashlight at a half-silvered mirror. The beam consists of untold numbers of photons, but let's focus on just one of them. When this particle impinges on the mirror, it enters a superposition in which it both reflects off and passes through. If you monitor these two possibilities using light detectors, you will find the photon has a fifty-fifty chance of going one way or the other. The act of observation forces the particle to choose a path, according to the collapse rule of standard quantum theory.

Wigner imagined that an observer conducts this experiment while a second person, the "superobserver," stands outside the lab, unable to see what is going on. The superobserver gives the observer time to complete the task, enters the lab, and asks what she saw. The observer thinks the particle collapses when she measures it. From the superobserver's perspective, however, the particle collapses only when he "measures" it—that is, when he queries the observer. So for some period of time, their views diverge. During that time, the observer has seen a definite result—the photon either passed through or reflected off the mirror—while the superobserver has had to assume it remains in a superposition of both options. By extension, the superobserver must have assumed that the observer, too, was in a superposition of having seen both options—a Schrödinger-cat-like state of limbo. To him, she is just part of the experimental system and should follow the same laws as the particle. So it's not just that he can't see her result. He doesn't even think she has gotten one.

At least that's what quantum theory tells him. But the superobserver does not have to take it on faith that the observer is in a superposition. He can prove it. Recall from chapter 4 that a collapse is the one process in physics we know of that can't be undone—and so theory says that the superobserver, in not having observed the collapse, has what you might call a superpower. Because no irreversible collapse has occurred for him yet, he could undo the superposition and restore the setup to its original unambiguous state. In the case of the

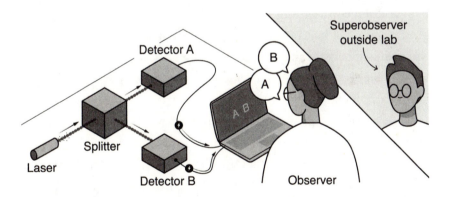

5.1. WIGNER'S-FRIEND THOUGHT EXPERIMENT. Consider the same quantum experiment we've been talking about all along: a photon takes both of two possible paths, and the apparatus registers that it took both. But now suppose the observer conducts the experiment behind closed doors, while a "superobserver" waits outside. The superobserver thinks the observer sees the photon take both paths—that she is effectively part of the apparatus. Furthermore, the superobserver could, in principle, act on the entire apparatus, including the observer, and reverse the experiment. Treating another person as a quantum system calls into question the nature of scientific objectivity.

photon, wielding this power is straightforward. Because the photon remains in a superposition of reflecting off and passing through the half-silvered mirror, the superobserver could add a couple of fully reflective mirrors to intercept the photon on both paths and redirect it back to a second half-silvered mirror. The mirror would meld together the waves corresponding to the two paths, reconstituting the original photon. In essence, the superobserver could force the photon to do a U-turn that would erase the ambiguity of the superposition that came before. It would be as if the photon never impinged on the half-silvered mirror in the first place.

Undoing the superposition would be more complicated when it's not just the particle but also the observer who is in a superposition; the superobserver would need to perform some insanely complicated operation on all the particles in the observer's brain. But quantum theory says it could be done. And then the observer would be back in the exact same condition she was in at the start of the experiment. She would see something, then unsee it, as though her memory had been wiped. For her, nothing would have happened. Some physicists and philosophers think such an experiment will become technically feasible before too long, with an AI system acting as a proxy for a human observer.[2]

The tension is not just that two people have differing views of the same events, but that the superobserver himself is torn. The reversal experiment would confirm that the observer is in a superposition, but other evidence suggests that she has seen a definite result. Namely, if the superobserver lets the experiment play out and asks the observer when the collapse occurred, she'll reply that it was when she measured the particle, and thus the matter was settled long before the superobserver walked in. If he insists she was in a superposition before he walked in, she'll complain he's gaslighting her. And if he still won't believe that she achieved a definite result, they can rerun the experiment, but this time she will slip him a note under the door as soon as she has made her measurement. "The friend can send a message outside of the sealed laboratory in a subtle way which will not reveal the information about the outcome, but just says, 'I see a definite outcome,'" Časlav Brukner said. As long as she states only that she saw a collapse, and does not divulge what the result was, her note will not count as an "observation"—and so, for the super-observer, no collapse will occur.[3] But he now holds in his hand a note that violates his previous understanding of the experiment.

NO COMMON GROUND

Disagreements are routine in science. Scientists have told me they're almost shocked when they find themselves in agreement. But the discrepancy that arises between the two experimenters in Wigner's setup is much worse because no amount of extra data or fresh thinking can resolve it. In 2015 Brukner demonstrated the depth of the problem by devising a clever elaboration on the original scenario.[4] It uses quantum entanglement as a probe. Usually in quantum experiments, physicists make observations to understand entanglement, but Brukner turned the tables and used entanglement to understand observations.

Brukner imagined setting up two teams of observers and superobservers, creating a pair of entangled particles, and giving one particle to each team. The observers measure their respective particles to ascertain whether the particles either pass through or reflect off a half-silvered mirror. Then the two superobservers probe their respective observers and compare their results. They repeat the procedure multiple times to look for statistical patterns. To mix things up, sometimes a superobserver simply asks the observer on his team what she saw. Other times, he exercises his superpower: he reverses the observer's experiment and wipes her memory, cutting her out of the picture so that he can measure the particle for himself.

This setup allowed Brukner to study what a memory wipe

means. On those occasions when a superobserver queries an observer, he learns that she obtained a definite result. On the occasions when he does the memory wipe, he doesn't know directly what she saw, but he can take a guess based on the statistical patterns he observes. If she obtained a definite result, that would break the entanglement, and the patterns would deviate from those of quantum theory. He sees no such deviation. Ergo, she did not obtain a definite result. Now, this is creepy. The memory wipe is acting not just as a memory wipe, but as a reality wipe. Her result is not merely forgotten, it never existed as far as the superobservers are concerned.

In 2019 a team at Heriot-Watt University in Edinburgh did the experiment using entangled photons and confirmed that a contradiction arises.[5] Since wiping a real person's memory is unrealistic, not to mention unethical, the experimenters brought in other photons to play the role of human observers. These extra photons "observed" the original photons by interacting with them inside a special optical element. Because the experiment presumed that a photon is an adequate substitute for a human—that both human and photon obey the rules of quantum theory equally—it was just a demonstration as opposed to a proper test.

As ingenious as Brukner's setup was, entanglement is a puzzle in its own right, so using entangled particles to prove a different point has a house-of-cards quality to it. The quantum physicist and philosopher Eric Cavalcanti at Griffith University in Brisbane, Australia, and his colleagues followed up with a more tailored analysis that closed some of the loopholes.[6] But that still left two concerns that can't be entirely assuaged.

First, perhaps it's not possible to conduct a randomized, controlled experiment in quantum physics. In Brukner's scenario, the superobservers decide at random whether to query the observer or wipe her memory. What if this decision is not actually random but is related, in some hidden way, to the observed outcomes? This putative bias, known as superdeterminism, could skew the findings and lead the superobservers to draw unwarranted conclusions. Physicists have checked for such an effect in other experiments. In one, they attached a telescope to their detectors, pointed it at distant galaxies, and used flickers in the galactic light to randomize their experimental choices. The galaxies were so far away that their light had been traveling for billions of years and was therefore independent of conditions on Earth, ensuring that the measurement choice was unbiased.[7] That didn't entirely rule out superdeterminism, but if it were still somehow true, we'd have a much bigger problem than an inconsistency in one particular experiment. If we can't trust our experimental protocols no matter how scrupulous we are, we might as well give up on science altogether.

A second loophole in Brukner's conclusion is that entangled particles may be linked by some hitherto unknown force that acts at a distance. What happens to one may instantly affect the other, even if they are kept well apart. (Many books and articles on physics *equate* entanglement with action at a distance. But this is a misconception that physicists and science writers, in our struggles to get across a very subtle idea, have inadvertently promulgated. In fact, entanglement is a phenomenon with multiple possible explanations, only one of which is action at a distance.) If so, the particles could flout the laws of chance by their own interaction, and the discrepancies would say nothing about the capacity of observers to reach agreement. But because such a force would operate instantly, it would define an absolute notion of time, violating Einstein's special theory of relativity.[8] Some physicists and philosophers are fine with that, but most conclude that the particle correlations must have some other explanation.

Based on these experiments, if you set aside superdeterminism and action at a distance, and if you think that quantum theory is as valid for observers as for particles, you are forced to conclude that observers can reach irreconcilable conclusions, leaving a troubling subjectivity in our most basic physics. It has long been a precept of quantum mechanics that experiments that are not performed have no results, not just in the literal sense that you never know until you look, but in the deeper sense that a particle does not even possess the measured property until then.[9] Now it seems that experiments that are not performed *by you* have no results for you, either.[10] "Unless *you* observe this, it should not exist," Brukner said. "Even if other observers have seen a definite outcome, you should not take it as such."

It's one thing for observers to disagree on the meaning of theoretical abstractions such as the wave function. It's quite another for them to disagree on the data they collect from experiments. "It's a bit disconcerting," said Cavalcanti's coauthor and colleague Nora Tischler. "A measurement outcome is what science is based on. If somehow that's not absolute, it's hard to imagine."

I KNOW WHAT YOU KNOW

As if that were not enough, Wigner's experiment comes in yet another version, and this one caused the biggest stir of all. Proposed in 2016 by the quantum information theorists Daniela Frauchiger and Renato Renner, both then at the Swiss Federal Institute of Technology in Zürich, it has been the subject of entire conference sessions and dozens of papers.[11] "I literally received thousands of emails regarding this article," Renner told me. Employing a different style of argument, he and Frauchiger showed that the meta-experiment is more than an academic disagreement about what observers think of one another. It leads to an outright contradiction between theoretical predictions and experimental outcomes.

Their experiment involved a convoluted series of measurements and logical deductions. This made the argument watertight but also hard to follow, which was one reason it got so many people's blood boiling. I have a 2,700-word blog post with formulas if you want the gory details.[12] Fortunately, the basic idea is simple. As in Brukner's experiment, you create a pair of entangled photons and give them to two observer–superobserver teams. The observers measure the particles as if flipping coins. Between them they can obtain four possible outcomes: both heads, both tails, one heads and the other tails, and vice versa. The superobservers, in turn, measure the observers, and their outcomes also look like coin tosses.

Frauchiger and Renner arranged things so that ordinarily only three of the outcomes are possible; only on certain occasions might you see the fourth. This slight restriction means that one observer or superobserver can sometimes know for sure what another will see. For instance, suppose all the outcomes can occur except for two heads. Then, if you see a coin land on heads, you can immediately deduce its entangled partner must have landed on tails, even if you never see it do so.

With this setup, the experiment is like playing a murder-mystery game such as Clue. The trick is to leverage your partial information by watching what other players do. At first, you might rule out everyone except Professor Plum or Miss Scarlet. Then, based on the questions people ask, you deduce that someone is holding the Miss Scarlet card, leaving the professor as the bad guy. Likewise, observers and superobservers in the experiment have some partial information about the particles. They take turns performing observations and make guesses about what everyone else is seeing. By chaining together these inferences, somebody might reach a conclusion such as "I know that you know that Bob knows that his coin toss is tails."

The person who succeeds in piecing together these inferences concludes that, when the time comes to perform his own coin flip, he should see, say, heads. Yet sometimes this person will see tails. So there's a contradiction. As in earlier versions of Wigner's setup, the person's logic fails because it assumes the other participants have obtained definite outcomes—a natural assumption, but one that it turns out he can't make. Rather, he has to deny what his colleagues take to be settled fact and instead track the superpositions that they enter into and consider how these superpositions can either offset or reinforce one other.

I performed the first experimental demonstration of Frauchiger and Renner's thought experiment in 2019 using an IBM cloud-based quantum computer, which used superconducting quantum devices to stand in for human observers.[13] Mimicking the logic that the "observers" would apply to make a prediction, the algorithm confirmed that the prediction was wrong about half the time.

As in the other variants of Wigner's experiment, the superobservers wipe the observers' memory. Not only is this discomfiting, many skeptics of Frauchiger and Renner's experiment also worried it invalidated the analysis. If someone has made an observation and then unmade it, are you entitled to base any conclusions on what they had seen? Fortunately, Frauchiger and Renner staged their experiment to skirt this objection. Observers get their memory wiped only after they have played their part in the logic. By then, whatever they saw and inferred has already been incorporated into the analysis, and nobody refers to it again. "This is somehow the main trick how we get around the main dilemma of Wigner's original thought experiment," Renner said. He and Frauchiger also closed off other outs, including superdeterminism and action at a distance. There are only two ways to escape their maze: either quantum mechanics breaks down at the level of observers or physicists need to give up objectivity.

By showing how observers can go astray in their Clue-style piecing together of information, the experiment casts doubt on a core principle of objectivity: that knowledge is transitive, which philosophers call "closure." By this principle, if your friend sees something and tells you, or you can otherwise deduce what she saw, that's just as good as seeing it for yourself. You can combine her observation with others to form a shared corpus of knowledge. Even if your friend comes down with a case of amnesia, the original information remains valid. Without this principle, we'd be lost: all our knowledge is a chain of inference. "No measurement we carry out is ever really direct, but rather mediated by others," Renner said. But now it seems that not all information can be neatly fitted together. Under some circumstances, there is no such thing as an absolute fact, one that is as true for me as it is for you. "There does not need to be a God's view that includes everything—all observations by all observers," he said.

IS REALITY A MATTER OF PERSPECTIVE?

Many physicists and philosophers embrace this relativism. For them, the deep lesson of quantum physics is that reality is observer-dependent, or "perspectival"; when you make a statement about reality, you must always specify *whose* reality.

That truth is observer-dependent is an old philosophical idea. Kant compared it to the Copernican revolution.[14] Earth is not the center of the universe; in fact, nothing is; and just as no position in the universe is privileged, no observer's vantage point is privileged over any other, he argued. The analogy to Copernicus carries an important lesson. As we watch the sun and stars rise and set in the sky, it looks as if we are at the center of the universe. But we shouldn't read too much into first impressions—it will *always* look to us as if we are at the center of the universe, whether we are or not.[15] We are stuck to Earth's surface, so we have no direct way of sensing our motion around the sun; we have to infer it from subtler clues, such as how the other planets occasionally appear to switch direction in the sky. Similarly, in quantum physics we seem to be establishing absolute facts, but can we really tell them apart from relative ones?

Perspectivalism entered quantum physics in the 1920s with Niels Bohr and his concept of "complementarity," by which he meant that some aspects of a particle are mutually exclusive. You can measure a particle's position vis-à-vis a certain direction or its speed in that direction, but not both at once, at least not with absolute precision. So what you know about the particle depends on your choice of what to measure. Bohr went further and argued that what the particle *is* also depends on your choice. You and the particle form a unified system. Measurement results are properties not of the particle per se, but of the both of you.[16] You and I may see very different things and be unable to stitch our findings together. David Bohm later described it as a twist on the elephant-and-blind-men parable, where one blind man feels a rope, another a tree trunk, a third a wall. But when they report their impressions to the king, he can't combine them into anything coherent, let alone an animal.[17]

Bohr pointed out that physicists are not total strangers to relativism.[18] Whether a car is zipping through the countryside or the countryside is rushing past the car is a matter of perspective. Physicists never talk of "the" speed of an object, only its speed relative to the ground, to the car, to the clouds, to the sun, or to anything else you like. Other measurements involving space and time also depend on your reference point, as Einstein spelled out in his theories of relativity. A car driver might see two events occur at the same time, while a pedestrian on the roadside thinks one preceded the other— and both are right.

Yet relativity theory does not deny there is some absolute shared reality. It just says that speed and simultaneity aren't part of it. It holds that although observers may disagree on the sequence of certain events, they will always agree on which event causes which.

Quantum physics takes a pickaxe to this bedrock of agreement. The disputes over clock time in relativity theory are piddling compared to what happens with Wigner's friend, where one observer denies that the other saw something happen at all. "Quantum mechanics is the discovery that facts are contextual," said Carlo Rovelli at France's Aix-Marseille University, one of the physicists who has been developing a perspectival view of quantum theory. "They are relative to physical systems. Facts are relative in the same sense in which velocity is relative: velocity is a property of an object relative to another object."

This doesn't mean that anything goes, or that reality is all in our heads. It just means that some types of observations can't be reconciled. "That things are relational or perspectival does not mean that they are not real," said the philosopher Dennis Dieks at Utrecht University. Furthermore, it takes an unusual situation, in which one observer performs experiments on another, to expose a conflict. As serious a problem as these discrepancies are for understanding nature at its roots, they don't mean there aren't empirical facts about climate change and presidential election results.

EVERYTHING EXISTS AT ONCE

Hugh Everett did the groundbreaking analysis of perspectival quantum theory in his PhD thesis in 1956. He then ran into a problem far more intractable than quantum physics: human arrogance. Bohr by this point denied that there was any measurement problem to solve. Despite the best efforts of Everett's graduate advisor, John Wheeler, Bohr and his disciples basically canceled Everett for his temerity in questioning their views. Everett left physics and went to work for the US military, tasked with developing nuclear war strategies.[19] (He was the one who figured out that the only winning move is not to play.) An irony is that Everett saw his perspectivalist views not as a repudiation of Bohr's principle of complementarity, but as a generalization of it.[20]

Using the same reasoning as Copernicus and Kant, Everett argued that physicists had committed a basic logical fallacy. The quantum wave equation predicts that objects enter a superposition of conflicting possibilities, but we never see them in one, so physicists figured they must collapse. They didn't stop to ask whether we *could* see such a superposition. Unless we can, our nonobservation of these superpositions tells us nothing. Absence of evidence is not evidence of absence. In fact, Everett went on, we can't see an object in two mutually exclusive states. The reason is that, according to the rules of quantum mechanics, superpositions infect us and we become part of the system we are trying to study, leaving us unable to see its overall state.

To explain how this happens, Everett imagined a bare-bones observer—not much to it, just an eye and a memory.[21] The eye watches a measuring apparatus, and the memory records what it sees. The memory can be abstractly represented as a quantum bit, or qubit, with value 0 or 1. For instance, if a photon reflects off a mirror, it takes a certain path and hits a detector. The eye notes this and stores a 0 in memory, representing reflection. If the photon instead passes through a piece of glass, it takes a different path and hits a second detector. The eye duly notes this and stores a 1 in memory, representing transmission.

Now suppose the observer watches a photon strike a half-silvered mirror. The photon both reflects off and passes through, and the observer's memory ends up in a superposition of 0 (representing reflection) and 1 (representing transmission), accurately reflecting the condition of the photon. In essence, the photon passes through and the observer registers that it passes through, *and* the photon reflects off and the observer registers that, too. There is no such thing as "the" path of the photon or "the" perception of the observer. The photon and observer are in limbo. But they are in it together—that's the key. The whole purpose of a measuring apparatus is to establish a correlation between the exterior world and our perception. If the world is X, we perceive X. So if the world is in a superposition, our perception must be in a superposition, too. And because the two superpositions are linked, the observer has a well-defined state relative to the photon. That is, each option in the photon superposition (pass through or reflect off) is matched with a corresponding option in the observer superposition (see it pass or see it reflect).

Furthermore, if you ask the observer whether it saw a definite result, it will look into its memory, find a value stored there, and answer, "Yes." The observer is never presented with a discrepancy—seeing the photon both transmitted and reflected—so it has no inkling of its own conflicted condition. This is perhaps not so unexpected. We humans always have difficulty getting perspective on ourselves— otherwise we wouldn't need therapists. Everett was suggesting that this lack of self-awareness enters into our most basic observations and is an unavoidable part of being embedded in a quantum world.

This is a classic case of the inside/outside problem. From the outside, both particle and observer are in a superposition of multiple measurement outcomes. From the inside, all the observer perceives is a single outcome. In other words, collapse occurs only on the inside. The observer sees a particle change from a superposition to a single outcome not because the particle itself has changed, but because the observer has become entangled with the particle and lost an outsider's perspective.

This is all we need to explain the various features of collapse, according to Everett's analysis. For instance, according to standard quantum theory, collapse is irreversible; Everett said it was enough that collapse *look* irreversible. From the outside perspective, it could, in fact, be undone, although that would require acting on all the particles in the superposition, and for more than a few particles, that's hard, bordering on impossible.[22]

In addition, thinking of collapse as the product of an insider viewpoint accounts for the random outcomes we observe. Everett showed this by analyzing a series of measurements on particles that were all prepared the same way.[23] The observer measures the first particle and stores the result. The memory is now in a superposition of seeing the particle go through and seeing it reflect. Then the observer measures the second particle and stores that result, too. Now its memory is in a superposition of four permutations: having seen the first particle go through and the second go through, having seen the first go through and the second reflect, and so on. On the next measurement, it has eight permutations to track, then sixteen, and so on. The overall state is entirely predictable—a ginormous tree of permutations—but the observer can't see that. All it sees is one series of outcomes, which, for a typical observer, will look like a toss-up. Much debate has ensued over what "typical" means,[24] but the basic point is that the notorious randomness of quantum physics is entirely mental, occurring because observers are stuck inside superpositions. In Everett's interpretation, randomness is a statement about us, not about particles.

Everett's way of thinking makes short work of Wigner's experiment.[25] Suppose you observe the observer. To you, it is in a superposition. If you ask it whether it saw an outcome, it will query its memory, find a value there, and reply, "Of course, silly." The two of you have divergent perspectives, and that's fine, since the state of a system is relative to who is measuring it. But as soon as you ask the observer what the outcome was—in essence, you use that observer as your own measuring device—its superposition will infect you. Now you will find yourself in the same position as the observer. You will think you saw an outcome, so you will think the observer and photon have collapsed. You will have given up your outside view for the inside view.

This analysis has a powerful appeal for physicists and philosophers because it dispenses with the collapse rule of textbook quantum theory. Columbia University's David Albert, who trained as a theoretical physicist and is now a leading philosopher of quantum theory, recalled encountering Everett's interpretation in the 1980s. "I had the most powerful conviction: this is so beautiful, this must be true," he told me. Yet he came to realize that Everett's presentation can't be the whole story. For one thing, it has a peculiar, self-negating quality to it: It says we don't see mutually incompatible outcomes because we're deceived about our own state of mind.[26] We think we've seen a single definite outcome when we're really in a superposition of having seen every possible outcome. For Albert and others, that sort of self-deception makes *The Matrix* or brain-in-a-vat scenarios seem tame by comparison. It's one thing to consider that we might not be seeing the world as it really is, quite another to entertain the possibility that we aren't thinking what we think we're thinking. "I could be hallucinating that there's a chair in the room, but I can't be hallucinating that I *think* there's a chair in the room," Albert said. If we were so profoundly deceived, we wouldn't be able to trust anything, including the observations and reasoning that led to quantum theory. So Everett's interpretation pulls the rug out from under itself.

Traditionally we expect scientific theories to satisfy two criteria: they should hang together, and they should match reality. By formulating theories mathematically, we can confirm they are internally coherent, and we can extract numerical predictions to compare with data. But there is a lesser-known third criterion: theories must not deny the validity of observations. A theory can be scrupulously logical and predictive, but if it covers its own tracks, then it fails the standards of science. It is, as philosophers say, empirically incoherent.

To check whether a theory negates itself, we need to insert ourselves into it—we need to ask whether it lets us test it. Albert and others argued that Everett's interpretation, in its original form, doesn't. Superdeterminism—the claim that we are unable to conduct a randomized, controlled experiment in quantum physics—comes close to empirical incoherence, too, but at least there are some ways to test for it. The next time someone tells you about a conspiracy theory, check for empirical incoherence. Sure, maybe cannibalistic pedophiles have taken over the government, and if ever you try to look for them, they stop you (or worse). It's logically possible; it would explain certain observations. But if you believe such a theory, why believe anything? Aren't you worried the conspiracy is to make you believe there is a conspiracy?

MANY MANYS

Everett left these loose ends because, for him, explaining our observations was enough. He didn't care much about questions concerning what was really going on, what the waves and superpositions in the equations correspond to in reality.[27] But as his views spread from 1970 on—becoming arguably the leading interpretation of quantum mechanics today—physicists and philosophers came up with various ways to plug this interpretive gap.

The best-known is the many-worlds interpretation. (Sometimes this term is applied to Everett's original work, but it really dates to 1973.)[28] The idea is that the universe splits into parallel universes. If a photon is superposed between passing through and reflecting off a mirror, you can think of it as two photons playing out both possibilities. Those photons ripple outward into the universe, and before long there aren't just two photons, but two of everything.

Essentially, this interpretation equates superposition with multiplicity.[29] A superposition of two options means two things exist out there. An observer sees one of them, and his doppelgänger in a parallel universe sees the other. Now there is no self-deception—each observer sees his world as it really is—but the reality we perceive is still relative to us.

A lot gets swept under the rug when physicists talk about "worlds." Quantum theory itself doesn't provide any guidance for how we're supposed to divvy up superpositions into "worlds." This is a consequence of the menu problem that I mentioned in chapter 4. In the case of the half-silvered mirror, we routinely talk as if the photon chooses from a menu of two options—it either reflects off or passes through the glass—and therefore there must be two worlds. In fact, there's a literal infinity of other, surreal menus in which the photon is reflected to varying degrees. The definition of a "world" is an add-on—some extra feature of physics above and beyond quantum theory itself. Physicists and philosophers have put considerable effort into articulating what that could be. Most now think it has to do with the process of decoherence. But this issue remains contentious. The idea of parallel universes, forming what is often called a multiverse, faces other challenges that I will explore in chapter 6.

For a while in quantum physics, interpretations that began with the word "many" proliferated almost as fast as the universes they claimed to describe: many threads, many spaces, many histories, many maps.[30] In 1988 Albert and colleagues developed one, the many-minds interpretation, in which it's not the whole world that splits, just your mind.[31] If you see a photon pass through the mirror glass and you also see it reflect off, there are two yous occupying the same brain and body. Your brain enters a superposition that amounts to two independent streams of consciousness. Those two selves go on to have other distinct experiences and will almost certainly never reunite. Put simply, all of us have a kind of multiple-personality syndrome.

The many-minds interpretation didn't go very far. Even its originators thought it was weird. "It didn't seem to us even at the time that anyone should really be prepared to accept it," Albert told me. One drawback was that it assumed that minds have continuity over time—that the mental state at one moment can be identified with a corresponding mental state at another.[32] Physics itself does not establish this continuity; it must be separately postulated. "Mind really is being treated metaphysically, ontologically, as a distinct object from, say, the brain or from anything in the physical world," Albert said of his interpretation. The continuity of identity is a major puzzle that I will return to in chapter 6.

Still, the many-minds interpretation is historically interesting because it brought neuroscience and philosophy of mind into the physics conversation. In this interpretation, the options we see in our experiments—such as a photon either reflecting off or passing through a mirror—have nothing to do with parallel worlds; rather, they originate in the structure of our minds. There's something about human thought that carves the world into options such as "reflect" and "pass through," as opposed to all the other possible dichotomies. To understand why that is, physicists need a theory of consciousness—speculating about consciousness is no longer just a fun diversion for them, but an essential part of understanding experimental results.

A WORLD OF RELATIONSHIPS

The core idea underpinning the many many interpretations is that we have access to only a very small fraction of reality. Somewhere beyond our view are worlds or minds by the billions. Only a god, standing outside the temporal realm, could view them all. Still, they exist independently of us or anyone else. Carlo Rovelli, for one, aims to dispense with this last vestige of absolute reality by taking Everett's approach in an even more thoroughly perspectival direction.

I became engrossed by his take on quantum mechanics when I was writing my first book in the mid-2000s. We had a long email exchange, and I visited him in the French Mediterranean village of Cassis. We went hiking along the rocky shoreline, first chatting about the pre-Socratic Greek philosopher Anaximander (on whom he was writing a book at the time), about our experiences of living in different countries, and about the difficulty of coping with uncertainty in science and in life. Only then did we get to the questions that had brought me there. For Rovelli, science is never just transactional. It is the establishment of relationships, both between people and between us and nature.

Throughout his career Rovelli has argued that the physical world, too, is a web of relations. He takes the essential lesson of modern theories of gravity and other forces to be that things have no properties in isolation, but acquire them only at their point of contact with other things. He extended the principle to quantum mechanics in 1996.[33] Measurements, he argued, are relations that we establish with something.

Although physicists frame quantum measurement in terms of observers, Rovelli doesn't think sentient beings have a fundamental role in reality. Their minds definitely don't cause wave functions to collapse. To him, they are just one type of physical system to which the quantities of physics can be related; a table lamp would do just as well. His view makes quantum theory completely egalitarian. If we measure a particle, we establish a relationship with it; if a lamp interacts with the particle, it establishes its own relationship with it. Thus the lamp is no less of an "observer" than we are, and its relationship is a "measurement" of sorts. "When I say that things are true relative to a system O, this has nothing to do with O having a mind," Rovelli told me.

Relations are necessarily specific to the involved parties. If you have a relation with something, and I have a relation with that same thing, and our experiences differ, no problem. You have your reality, I have mine, and we will have to agree to disagree. "If quantum mechanics is correct, every time we give a full description of what we think is the 'reality' of a situation, we are in fact only giving a partial picture," he said. "It is a 'reality as far as we are concerned.'" This is how he makes sense of the diverging views in Wigner's experiment.

Rovelli's relational interpretation of quantum mechanics has a purity and evenhandedness that leads to strange and, to skeptics, implausible conclusions. When he says that reality is strictly relational, he means it: when a particle is just on its own, not interacting with a person or a table lamp, it has no properties, period. The quantum wave function does not describe the condition of a particle at these intermediate times, but only the correlations that will occur once it finally does interact with something else. Not only does a tree falling in the forest with no one to hear it make no sound, but it doesn't even exist. Rovelli said: "In the relational perspective, you are not supposed to ask, 'What is the real state of affairs?' but only, 'How will an object manifest itself next?'" In between these interactions, there's a whole lot of nothing. Our observations are not snapshots of a world that existed before we came along and carries on existing afterward. They are all there is. That means reality has a staccato existence; it winks in and out. Dennis Dieks, though broadly sympathetic to Rovelli's approach, finds his picture of an intermittently existing world "rather outlandish." Rovelli doesn't dispute that it's strange. "The resulting world is still weird, very much so," he told me.

Philosophers have pointed out that Rovelli's approach has much in common with a school of thought known as structural realism, which in its purest form holds that physical things are nothing but their relations; they have no intrinsic qualities.[34] By analogy, consider a famous optical illusion by the German psychologist Walter Ehrenstein, in which lines converge on a point but never actually meet, like a wheel with spokes but no hub. We still see a hub, because the brain fills it in. Similarly, we see relations in the world, and our brains presume those relations must be anchored in concrete objects, but maybe those objects are illusory. Advocates think this interpretation dissolves the hard problem of matter—the puzzle I mentioned in chapter 1 that the laws of physics describe how things relate, but not what they are. There is no need to worry about what things "are" if they don't exist.

Structural realism makes reality sound like a caricature of Los Angeles. You think you're driving from one place to another, but really you're just endlessly circulating on freeways, never arriving because there's nowhere to arrive at. Skeptics of structural realism argue that relations without objects are as meaningless as freeways without towns. A relation alone is a mathematical abstraction; what breathes fire into it, they say, is that there are physical things on either end.[35] Skeptics also worry that the theory is circular.[36] "If something is defined by what it does, then it's defined in terms of its impact on something else," said Philip Goff, a philosopher at Durham University. "But then that thing will be defined in terms of its impact on something else, and that thing . . . and so on, ad infinitum. So you end up either going on forever or in a circle."

PRAGMATISM OF THE PHILOSOPHICAL KIND

Other perspectivalist views back even further away from thinking of the waves and superpositions of quantum physics as structures that really exist. Maybe quantum theory is not a direct representation of reality, even relationally. The much-fretted-over paradoxes could be an artifact of taking the equations too literally. Those equations may say more about our own ways of thinking than about the world itself.

"I think a good bit of the problem comes from something that was beat into most of us at an early age," the physicist Chris Fuchs told me in 2004, in the early days of formulating his own interpretation. "It is this idea: Whatever else it is, quantum theory should be construed as a theory of the world. The formalism and the terms within the formalism somehow reflect what is out there in the world. Thus, if there is more to the world than quantum theory holds out for, the theory must be incomplete, and we should seek to find what will complete it. But my tack has been to say that that is a false image or a false expectation. Quantum theory from my view is not so much a law of nature—as the usual view takes—but rather a law of thought."

Fuchs traced his thinking to a class he took at the University of Texas with Wheeler. "In the spring of 1984, John Wheeler said to a student standing beside me, 'I am well prepared to believe that an electron is nothing more than our information about it,'" Fuchs recalled. "I wrote it down in my notebook." Fuchs's notebooks are the stuff of legend. He occasionally posts all his email correspondence online, with the warning "Do not unthinkingly print from a printer," because it runs to thousands of pages.[37]

Fuchs is now at the University of Massachusetts Boston, and his view is now known as QBism, a double entendre of the sort that physicists are especially drawn to. First, it stood for quantum Bayesianism (after Thomas Bayes, the eighteenth-century mathematician we met in chapter 3), then for quantum Brunism (after Bruno de Finetti, a twentieth-century mathematician who studied the meaning of probability), then for quantum bettabilitarianism (since it involves placing bets, at least conceptually). But none of the Bs rang quite true to Fuchs, so now the acronym doesn't stand for anything in particular. It is pronounced "cubism," because its whole point is that quantum mechanics is not representational.

Many share that view. "I don't think that quantum theory is in the business of providing an answer to the question of what is the world made of," agreed Richard Healey, a philosopher of physics at the University of Arizona. He calls his own version of this approach a "pragmatist" interpretation, following a line of thinking that goes back to William James, the nineteenth-century philosopher of mind.[38] To judge the truth of something, James argued, you don't need to trace it all the way back to its ultimate nature—what philosophers call its ontology. That may be inaccessible to you and not directly relevant, anyway. What is important is what works. "Quantum theory works so well because it gets the right probabilities, not because it explains why these probabilities are correct," Healey said.

Much of our thinking in daily life is pragmatic. Do you honestly know why a toilet flushes when you pull the handle? Stock traders get rich on watching for upticks and downticks in the markets without a clue as to what the companies they buy or sell actually do. Pragmatism also resonates with ideas in machine learning: The whole point of a machine-learning system is to analyze data and make predictions, not necessarily to capture the world as it really is. A neural network that classifies images as puppies or kittens may have no conception of pets. It might work off of combinations of pixels that to us look like visual snow.

"The machine-learning approach to science is not to find the

underlying mechanisms in reality that caused the phenomena to occur, but rather to predict," said Elad Hazan, a computer scientist at
Princeton. "It doesn't have to be meaningful to the human." He
gave the example of controlling a robot arm. Instead of working out
all the complicated physics of its movements, engineers can set up the
controller to learn by doing. The solution still has to obey the laws
of physics, but it might be a complicated mash-up of the physics that
bears no resemblance to the laws of motion as we normally formulate them. "You create a model which doesn't correspond exactly to
the physical reality of the arm, but can still control the arm," he said.
If you always insisted on comprehensibility, you'd deprive yourself
of a good robot controller, and perhaps much else.

We expected more from quantum physics, but Healey takes the
view that it is what it is. "We shouldn't insist that a quantum theory
come up with its [own] ontology and desperately look for one," he
said. "I think that's a wild goose chase."

Fuchs attributes the interpretive difficulty to the role of observers. He notes that an observation is never passive: in making one,
we establish contact with something in the world around us, and
the interaction causes an alteration in us. Whatever information we
acquire would never have existed had we not taken the initiative.
Thus we do not uncover facts, but create them. "Doesn't that just
make you tingle?" he said. "That—metaphorically, or maybe not so
metaphorically—the big bang is, in part, right here all around us?
And that the actions we take are *part* of that creation!"

On Fuchs's view, we can never hope to achieve a fully third-
person perspective because we are unable to separate ourselves from
what we study. Our own actions have feedback effects that we can't
analyze from first principles. In war, no plan survives contact with
the enemy—the situation is just too complicated and fluid. And so
it may be with quantum systems. What we *can* do, though, is to ex-
trapolate from past observations to predict future observations. Different observers have different sources of information and therefore

draw different inferences. The equations of quantum theory are a generic system for collating this information in an optimal way. It's not the only conceivable method of reasoning, but it happens to be the one that works for the physics of our world, so in an indirect way it gives us clues as to what is going on at a deeper level, Fuchs hopes.

In the nearly two decades I've known Fuchs, he has been trying to piece those clues together, and lately he has been leaning toward a relational view of nature, albeit in a different sense from Rovelli's.[39] But this side of his project is still very hazy. For now, Fuchs is much clearer on what quantum theory says about the observer than what it may reveal about the external world. And until he does articulate such a deeper view, how can we be sure QBism really dissolves the paradoxes of quantum theory?

Eric Cavalcanti in Brisbane said his work on Wigner's experiment has made him sympathetic to QBism, but he isn't entirely happy with it: "I still feel a certain sense of unease with this idea, as if walking over a hard glass floor. . . . A pragmatist account will definitely not satisfy everyone! Certainly not anyone who thinks of science as the search for knowing what the world is really made of."

Science is imperfect, goodness knows, but at least we expect it to tell us something about reality. For example, if a theory of electricity is formulated in terms of electrons, we take electrons to be real; maybe we will change our minds one day, but for now that's our best guess. A pragmatist view of quantum theory would leave us completely adrift as to what the world is made of. And it may come to that in the end. But most physicists still hold to the intuition that quantum physics reveals a world beyond us, however strange it might seem.

WHAT MAKES OUR SHARED REALITY?

Starting from our innocent desire to understand how light can behave as both wave and particle, quantum mechanics has taken us down a deep rabbit hole, making us question not just the nature of reality, but our own capacity to grasp it. But in stressing that different people can reach radically different conclusions about the same events, physicists now must confront the opposite problem: Why do we ever agree on anything?

Normally we explain our shared experiences straightforwardly: There's a reality independent of us. You see it, I see it, and any disagreement we have is on us. With more data or careful thinking, we can reconcile our views. But those who advocate perspectival interpretations of quantum mechanics give up on an external standard of truth. If you see a particle veer left, and I see it veer right, our views may never align. Worse, this misalignment is oddly hidden from us. We carry our divergences around with us, yet when we meet in person and share notes, we find ourselves agreeing. Even in Frauchiger and Renner's experiment, a contradiction occurs only by way of a chain of reasoning involving what other people saw. In any one-on-one exchange, everyone is in agreement. "If they do communicate, then they will all always agree," Cavalcanti said. "They won't look at the same device and see different things." How strange that reality should be personal, and yet different observers' views almost always mesh with one another. Depending on your mindset, that is either a bug or a feature of these interpretations.

Advocates reckon that the mutual consistency of our observations need not be built into nature as a consequence of an objective reality common to us all, but can be derived from quantum physics. They say our observations agree because observers can be entangled as surely as particles are. Skeptics are dubious.[40] Both Rovelli's relational interpretation and QBism guarantee only the consistency of a single observer's measurements. If I measure a particle, and I measure you measuring the same particle, I'll get the same answer. But that is subtly different from saying that you and I have seen the same thing. By what standard could you judge that the things are the same? That would require a viewpoint beyond us, which both interpretations eschew.

Jacques Pienaar, a quantum physicist who works with Fuchs at the University of Massachusetts Boston, has studied the difficulties of achieving mutual consistency in these interpretations. "There is nothing to guarantee that observer A's representation of observer B's view is an accurate representation of the view that B actually holds," he said. In fact, you can get strange conspiratorial scenarios in which A agrees with B only because A wiped B's memory and overwrote her original result with his.[41] So these interpretations have the same self-negating quality as Everett's original proposal: they take quantum mechanics to be objective fact, yet deny the category of objective facts. A physicist turned philosopher who has made this point, Emily Adlam of the University of Western Ontario, has managed to bring Rovelli around. "I had to argue with him a long time," she said. The two have added nonrelational elements to his interpretation, so that observers can find some common ground.[42] Though a partisan of QBism, Pienaar thinks it, too, may need to be modified to ensure intersubjective agreement: "I think it is both possible and desirable for QBism to postulate shared measurement outcomes between agents. But this would have to be an additional postulate."

Jan Walleczek, a biologist at Phenoscience Laboratories in Berlin who studies the interface of quantum theory and consciousness, is struck by how the theory stands the hard problem of consciousness on its head. Scientists have long been baffled at how our subjective experience could flow from their objective theories, but maybe that gets the question exactly backward. He said: "To me, the puzzle always was: How is *objectivity* possible? That, to me, is a totally magical thing." Solving this puzzle might require an even more thorough inversion of our thinking, and hints in that direction have come from another branch of physical science: cosmology.

6

MINDING THE UNIVERSE

AFTER A LONG DAY OF listening to lectures, the quantum physicist Markus Müller and I retreated to the hotel bar. This was the day after my train ride with Karl Friston in 2018, and we were at a workshop with neuroscientists, philosophers of mind, and physicists. Most implored physicists to pitch in to help the other disciplines understand the mind. Müller was going to speak the next afternoon on how physicists need help, too. "There is obviously the hard problem of consciousness that people talk about," he told me. "They say, 'Isn't it remarkable that we have this actual experience, and how do we account for that physically?' Now, my claim is that this problem is just one problem in a multitude of problems in and around physics."

Like many other physicists who work at this disciplinary intersection, Müller traces his interest in the mind to a transformative personal experience. In the '90s Germany still had a military draft, but Müller, like many young men, opted to do the alternative civil service. He was placed in a special-needs school for visually and cognitively impaired children east of Nuremberg. One of his new charges was "Mia," age four. She barely ate or drank. "I was told back then she had only a few years left to live," Müller recalled. As able-bodied people often do, he assumed she would be wallowing in self-pity.

Instead he found a bubbly little girl. She had the will and found the way. "She couldn't move very well," Müller said. "She could move only one arm. On the floor, to play with some stuff, she would always just move in a circle." He realized that her resistance to eating and drinking was not pathology, but normality. What four-year-old willingly eats broccoli? So he made a game of it. "I would answer 'Ba!' And she would answer 'Ba ba ba!' and just play word games. And then after, like, a couple of minutes of word games, she would drink one more drip of water."

After completing his service, Müller occasionally visited the school and, a few years ago, came across Mia's obituary in a newspaper. Defying doctors' predictions, she had lived to twenty. His experiences with her and the other kids stayed with him as he studied physics. "This really made me think about, Who are we? What does it mean to be human? What does it mean to be in the world?" he told me. These humanistic concerns may seem out of place in physics, but he thinks they are ultimately the reason we do physics at all.

Working at the Institute for Quantum Optics and Quantum Information in Vienna, Müller delves into what quantum mechanics reveals about reality. "These insights challenge our naïve picture of a world out there in the usual sense," he said. He is also driven by the puzzles of cosmology. Scientists who study the universe on its grandest scales struggle with the inside/outside problem no less than those who specialize in the finest grains of matter. Both specialties want to reconcile third-person and first-person views of the world—the objective perspective that physics traditionally seeks and the experience of an embedded observer.

To unknot this tangle of puzzles, Müller has proposed reimagining physics from the inside out.[1] Instead of starting with a third-person description and asking how it gives rise to a first-person one, he starts

with a first-person viewpoint and sees how far he can get in recon-
structing a third-person view: "We could have fundamental laws
that act on the level of the observer directly and then the world can
be an emergent phenomenon," he said. Few go as far as he does, but
most cosmologists have come to accept that theories of the universe
are as much about us as they are about everything else.

SPACE IS BIG—REALLY BIG

Three developments in modern cosmology have created a
predicament related to our place in the universe. First, scientists are
realizing the universe is big. That may not come as a surprise, but
it's striking how each discovery makes it bigger still. A century ago,
astronomers thought our galaxy constituted the whole universe, implying a size of a few hundred thousand light-years.[2] Today, our
microwave instruments observe structures—dense patches of primordial hydrogen gas—that are 46 billion light-years away. Within
that distance are, by one estimate, 2 trillion galaxies.[3] The distance
has little to do with cosmology per se; it is defined more by geologi-
cal and evolutionary time scales. It took 13.8 billion years for our
galaxy and planet to form and humans to evolve, over which time
light traveled a certain distance and space expanded by a certain factor.[4] If we were having this conversation 5 billion years ago, we'd
be remarking that the universe was about 20 billion light-years in
radius. So we shouldn't take the observed size as anything other than
a lower bound on the total size of the universe.

And indeed there is no sign that the universe stops at the outer
limits of our vision: galaxies or their precursors fill space all the
way out. Nor is there any hint that space curves back on itself to
form a closed ball—no equivalent of ships sinking beneath the ho-
rizon and no repeating patterns in the sky as light loops around.
If space were curved in that way, parallel lines would eventually

meet, and triangles would no longer subtend 180 degrees. Scientists have looked for such geometric deviations using cosmic structures as naturally occurring rulers, and they see no sign of curvature to within the precision of their measurements. That does not rule out a ball-like geometry of space, but it does mean that the ball would have to be a hundred times bigger than the volume we observe.[5]

Although cosmologists can't exclude more complicated, pretzel-like shapes that would confound their geometric analysis and make space look larger than it actually is,[6] other lines of evidence also indicate immensity. In addition to looking straight out 46 billion light-years, you can look laterally across the sky. You might see a galaxy or a pregalactic cloud of gas near the outer range of your vision, and then swing your telescope toward another galaxy or its precursor at a similar distance. If those two galaxies are more than a certain distance from each other in our sky, they are so far apart that 13.8 billion years hasn't been enough time for light to cross the distance between them. In other words, we can see pairs of galaxies that can't see each other.

And mysteriously, any two such galaxies look pretty much the same. No light, matter, or force has been able to cross between them, so what could possibly have coordinated their appearance? The answer is still being debated, but most cosmologists think it has to do with the way the universe expands in size with time.[7] Right now, it is growing at a fairly sedate pace. But if it underwent a growth spurt long ago, galaxies that were once close enough to influence each other were then torn asunder. This process of cosmological or cosmic inflation would have made the universe crazy-big, and quite likely infinitely big.

You might think nothing in nature could ever be truly infinite, but inflation is a self-perpetuating process. Although it has ended in the region of space we observe out to 46 billion light-years, it may continue far beyond our view, like a forest fire that is stamped out in one place but keeps throwing out new hot embers. The cosmic growth spurt could go on forever; in fact, it may already have been occurring for an eternity, in which case 13.8 billion years is the age just of our own patch of space.[8] If so, the universe will stretch on without end. "It would be cozy if it were finite," said Anthony Aguirre, a cosmologist at the University of California, Santa Cruz. "But it seems to me that eternal inflation gives you an infinite universe, and something like eternal inflation is happening now and probably happened in the past. I feel like nature is rubbing infinity in our face."

MONKEYING WITH THE UNIVERSE

The second issue that has forced cosmologists to rethink our place in the universe is that physics is randomized. Galaxies coagulated from clouds of gas that were scattered willy-nilly. The laws of quantum physics introduced further randomness; a given situation can unfold in multiple ways, each with some probability. Over the immensity of space, nature rolls the dice countless times, so there is a monkeys-on-typewriters aspect to the universe.[9] Patterns occur and recur by the luck of the draw. Whatever can happen, does happen. The only question is where.

Then there is the third issue: not only are the scattering of matter and the outcomes of events random, so are the laws of nature. One clue is that these laws are filled with seemingly arbitrary quantities. The mass of the top quark is 338,600 times the mass of the electron. The strong nuclear force is 137.0360 times as strong as the electromagnetic force. The universe currently contains 2.2 times as much dark energy as matter. There doesn't seem to be any logic to these numbers.

In fact, the only pattern that physicists have been able to discern is that these values have something to do with us—"us" in this case meaning any complex structure, not humans specifically.

The quantities are such that they enable galaxies to coalesce, stars to shine, and the universe not to implode, which are all preconditions for any imaginable form of life. A deviation from the observed values would, in many cases, prevent all this from happening and render the universe uninhabitable.

Much has been written on these so-called anthropic coincidences, although cosmologists in recent years have backpedaled on how significant they really are.[10] Some quantities turn out to have a lot of wiggle room; they could vary substantially before snuffing out life. Quantities that do seem finely tuned might eventually turn out to have some logic to them that has nothing to do with us; particle physicists already have some tantalizing hints of that.[11]

But until they come up with a better idea, physicists provisionally assume the values are tuned to our existential requirements because of an observer selection effect, as mentioned in chapter 1. On this view, particle masses and other quantities are not fixed properties of nature, but variables that billions of years ago bobbed up and down like stock prices until they froze at their current values. The point at which they froze was random, and the outcome varied with location. Other parts of space got different values from ours, and many, perhaps most, were left barren, unable to support galaxies, stars, and sentient observers. The amount of dark energy was especially vital: it doesn't take much of it to destroy everything else.

On this view, space contains an archipelago of isolated islands of matter governed by different laws—not a "uni"-verse, but a multi-verse. We find ourselves on one of the happy islands simply because we must, much in the same way that we find ourselves on Earth rather than on one of the other planets of the solar system because those other planets are too hot, too cold, or just plain too miserable. Our observations are biased by the simple fact that we can't make them if we don't exist. Therefore, to explain the values we observe, cosmologists have to incorporate observers into their experimental predictions. The irony is that the bigger the universe is, the more our own nature matters to understanding it.

The situation is uncannily similar to what happens in quantum mechanics. As I discussed in chapter 5, many physicists, too, think the quantum world is a kind of multiverse: If a photon hitting a half-silvered mirror can pass through or reflect off it, there is a sense in which it does both, even if we see only one outcome. All that can happen, does happen. The main difference is that, in the quantum case, those multiple outcomes all occur within the same volume of space and are inaccessible not because they are far away, but because quantum waves can overlap without interacting. The waves corresponding to a transmitted photon and to a reflected photon could pass through each other like ghosts and not even know the other is there. For observers, though, the effect is the same. It doesn't matter whether alternative outcomes are hidden from us because they are vastly far away or because we are unable to interact with them.

THERE'S ANOTHER YOU OUT THERE

The vastness of the multiverse transforms how physicists think about reality. It's not enough to say what the laws of physics predict, since they predict that everything will happen somewhere. Rather, you need to know what they predict for *you*. If a coin is tossed and you ask, "Heads or tails?" physics answers, "Yes." It will be heads in some places, tails in others. Which you see depends on where and who you are.

The information about where you are located in the grand scheme of things is referred to as "indexical." In a multiverse, all information is ultimately indexical. Knowing that the coin toss came up heads doesn't tell you about the coin, since it also landed on tails elsewhere in the multiplicity of worlds. Rather, it restricts where you could be located. "It's striking to me that this weird kind of information, this indexical information that no one really thinks about very much, could be the only one that we have," Aguirre said.

Once you do begin to think about indexical information, it makes sense of mysteries both great and small. Consider the irritating fact that, when you're waiting in a line at a supermarket or tollbooth, the other lines always seem to move faster. It's not always the fault of the checkout clerk or tollbooth attendant in your line, nor is it just your bad luck. Sometimes it's an indexical effect: a slow line has more people in it, so that's where you're most likely to be.[12]

Where things get strange is that you don't necessarily have the indexical information that you need. You might think you know where and who you are, but are you sure? When space is big enough, not only does everything that can happen, happen, but it happens over and over. If the universe is big enough, the conditions that gave rise to Earth, to humans, and to you are replicated somewhere out there. There are copies of you out there—creatures that are identical to you in every way and answer to your name. Some of your cosmic doppelgängers are ever so slightly different, playing out every variation that is physically possible. On top of having to decide what the laws of physics say for you, you face the discomfiting fact that you are not even unique.

This is where physics meets my favorite topic in all of philosophy: personal identity. When you wake up in the morning, how can you be sure you are the same person you were last night? How does your brain with its multitudinous parts construct a single self? What happens when it doesn't, as in dissociative disorders? Philosophers through the ages have devised ingenious thought experiments to explore these issues. In an Enlightenment version of *Freaky Friday*, John Locke imagined a prince who finds himself inside a cobbler's body[13]—a dramatic example of the inside/outside problem. If the cobbler insists he is really a prince, does he get to move into the palace? Or is he just a cobbler in need of a good therapist? Locke also imagined copying a person's mind into a new body, thus creating two beings claiming to be the same person.[14]

A multiplicity of yous means that, when applying the laws of physics to predict what you will see, you must ask, Which you? The objective state of the universe is like a map without the all-important "You are here" arrow, because the laws of physics fail to tell you where to situate that arrow. "The failure is that the question 'What happens to you?' is not just a question about the physical world," Müller told me. "It's not about a question like 'Where will that particle be?' or 'What does the differential equation predict here and there?' It's a different kind of question."

THE PARADOX OF THE ABSENT-MINDED DRIVER

In the 1990s philosophers and economists became fascinated with what happens when we're missing indexical information and aren't sure where we are.[15] This uncertainty throws a wrench into the rational judgments we make, and equally plausible ways of reasoning can lead to conflicting conclusions. Some of these puzzles are baroque, but one that rings all too true is the paradox of the absent-minded driver.[16]

Suppose you're driving to your in-laws' house and can't get a cell signal, so you have to resort to the old-fashioned system of following directions. The road has two turnoffs. The first one is the long way, the second one a shortcut. If you miss both and keep going straight, you'll still make it; this third route is intermediate in distance between the other two. You'd obviously prefer to take the second exit, but you know you tend to lose track of where you are and aren't up to the task of counting turnoffs. So you are driving along, not sure where you are, and you reach an exit. Do you take it?

The paradox is that the optimal course of action depends on whether you're deciding in advance or in the moment. If you're planning ahead, you figure that if you choose to take an exit, you'll end up taking the first one you come to, which is the worst of the three routes. So you vow not to take any turnoffs at all; the main road is the best compromise. But once you hit the road and arrive at a turnoff, things look different. You think to yourself, I might have driven past a turnoff already and forgotten it, so this could be the shortcut—there's a fifty-fifty chance. Now it makes sense to exit.

Psychologists have given an analogous scenario to test subjects. By asking participants to make their decisions under time pressure while also performing other tasks, the researchers created a kind of situational absent-mindedness—like the famous gorilla experiment in which people were so intent on counting the number of basketball passes during a game that they missed the gorilla walking through the scene. Aware of their distractedness, participants did the equivalent of exit the road more often while executing the action than when planning it in advance.[17]

Physicists developing the many-worlds interpretation of quantum mechanics have gotten a lot of mileage out of scenarios like these because the driver's conflicting strategies turn out to be analogous to different ways of calculating probabilities in a cosmological or quantum multiverse.[18] Whenever you blank out and forget where you are, you are reproducing the experience of those identical yous in a multiverse, unsure who is who.

These researchers think the uncertainty of quantum physics is indexical. When a photon strikes a half-silvered mirror, the equations leave no doubt about what happens: the photon enters a superposition. But there is plenty of doubt about what you will observe. In the many-worlds interpretation of quantum theory we considered in chapter 5, that's because some copies of you will see the photon reflect off the mirror and some will see it pass through. You can't tell who's who, so you have to allow for both eventualities. The probabilities generated by the theory are for you, not for the photon.

"There is no true randomness in the cosmos, but things can appear random in the eye of the beholder," said the cosmologist Max Tegmark, an advocate of the many-worlds interpretation. He has worked with Aguirre on interpreting quantum uncertainty as indexical uncertainty.[19] "If you have any mechanism of cloning, observers will perceive objective randomness if they're cloned. . . . The randomness reflects your inability to self-locate," he told me. By this reasoning, you may never get to meet your doppelgängers, but you sense their presence every time something apparently random happens.

In short, questions of identity are not just philosophical navel-gazing. They have tangible consequences for science.

ATTACK OF THE BOLTZMANN BRAINS

In the absent-minded driver scenario and conundrums like it, you may not know where or who you are, but at least you're not being deliberately fooled. In other situations, you can't even be sure of that. Descartes worried that the world you experience might be a hallucination created by a malicious demon,[20] a speculation that spawned a long literature of brain-in-a-vat stories, most famously *The Matrix*.

Today these scenarios usually involve artificial intelligence. I asked the cognitive scientist Joscha Bach how hard it would be to duplicate your mind in a machine. "Actually, very easy," he replied. "It's necessary and sufficient to build a machine that thinks it's you." Bach subscribes to Locke's theory of identity, according to which you consider yourself the same person as yesterday not because you have the same body, but because you remember being that person. Memories make the man.[21] An advanced AI system could run a simulation of your brain and—according to at least some theories of consciousness—be just as sentient as you. It wouldn't be a full dupli-cate, because transferring all your memories to the machine without destroying you in the process is beyond any foreseeable technology. But you could seed the machine with enough memories to give it a sense of psychological continuity with you. "Your identity is only given by your memories telling you that you are the same person as yesterday," Bach told me. "That's all there is. If I can give an ar-bitrary system the memory that they were you yesterday, they will think that they are you."

In some brain-simulation thought experiments, future tech companies run not just one brain simulation, but hordes of them. They could create a dozen or more simulated yous versus the unique real you, so brace yourself: you are already more likely to be virtual than real.[22] The saving grace is that setting up vast data centers to run copies of twenty-first-century humans would be a lot of work even for a twenty-third-century Elon Musk. I trust he'll realize that we're really not worth it.

But cosmology predicts a naturally occurring version of the brain in a vat. If the universe lasts long enough, it will dissipate its energy and degenerate into a condition that cosmologists call heat death. Activity will not cease altogether, but will become fully disordered—in essence, what is done will be quickly undone. We're well on our way to heat death. Galaxy formation has basically ceased already, star formation is at a tenth of its peak rate, concentrated forms of energy such as nuclear reactions are being turned into de-graded forms such as heat, and, distressingly, even the fabric of space is showing wear and tear.[23] The last of these is a consequence of the accelerating expansion of the universe, which is reshaping space into what cosmologists call a de Sitter geometry—a kind of heat death for space.[24]

In heat death, the universe is a twitching corpse that will stir at a low level for all eternity. Particles will career around, occasionally coming together and quickly parting ways. They'll have plenty of

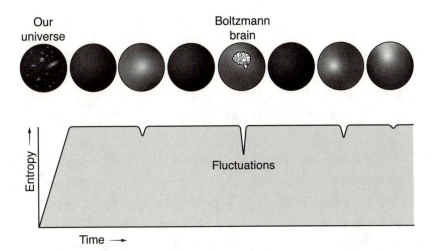

6.1. BOLTZMANN BRAINS. Once stars exhaust their fuel and the universe runs out of useful energy, it will approach a state of maximum entropy for as long as it continues to endure—possibly forever. Nothing will happen anymore, except for the occasional random fluctuation. On rare occasions, those fluctuations will produce a recognizable structure, even a conscious being—a so-called Boltzmann brain. Over eternity, an infinite number of such transient beings will arise. Many cosmological theories predict you are more likely to be one of them than a product of ordinary cosmic evolution.

chances to blunder into transient structures. Swirling clouds might even assume the form of a person—and not only the form, but an actual person, with thoughts, memories, and sensations. Such an apparition goes by the name of a Boltzmann brain, since it was Ludwig Boltzmann who first speculated about the ongoing activity that occurs even in heat death.[25]

It will be incredibly rare for a person to spontaneously materialize out of random particle collisions, but eternity is a long time. Sporadic though these Boltzmann brains may be, they will ultimately outnumber all the brains that ever formed the old-fashioned way. Numerically, you are likely to be one. In fact, in terms of the probabilities, you are likely to have formed a millisecond ago and to disperse a millisecond from now. All your memories of a long and happy life on a cozy planet are a lie. At any moment, the veil will lift. "The Boltzmann-brain problem would be, you think you're on this planet now, and suddenly you realize, Oh, I'm in space," Müller said. "I'm a random fluctuation. Waaaaaa! And you disappear."

GETTING RID OF THESE BRAINS

Cosmologists don't for a moment think we're actually Boltzmann brains—they're not quite that crazy. To the contrary, they take the prediction of these freak brains as a sign that their theories must have a screw loose. A theory based on observations had better not reach the conclusion that observations are fake.[26] In other words, Boltzmann brains are empirically incoherent. Maybe we are deceived about our own mental states, or maybe we are spasms in a universe long dead, but if so we might as well resign our professorships.

So cosmologists agree they have to get rid of Boltzmann brains, but they can't agree on how. "Just using the physics laws as you know them will not resolve the Boltzmann-brain problem," Müller said. This is because the mathematics of infinity makes it hard to assess the probability of these brains, or of anything else. If an infinite number of yous toss a coin, an infinite number will get heads, and an infinite number will get tails. So what are the odds that any one copy of yourself will see heads? That's the ratio of two infinite numbers, which doesn't have a unique answer. To try to measure the ratio, you could pair each self who sees heads with one who sees tails. That would imply their numbers are equal, giving fifty-fifty odds. Alternatively, you could form groups of three: a self who got heads with two selves who got tails, and because there are infinitely many selves, you can sustain this imbalance indefinitely. That would imply heads is half as likely as tails. Pick any group size you wish, and all are valid. This ambiguity means the laws of physics, on their own, no longer make concrete predictions for what a given observer will see, or even whether the observer is real.

Some scientists think that Boltzmann brains and other puzzles of probability mean they should give up on cosmological inflation and seek some other account of the universe's history.[27] Those who defend inflation draw a different conclusion. They think the theory of inflation can be supplemented with a "measure," a rule for how to count the members of an infinite set. This measure goes beyond the usual laws of physics, and one influential idea is that it has to do with our capacity to observe.

This proposal builds on one of science's foundational commandments, articulated well by Raphael Bousso, a physicist and cosmologist at the University of California, Berkeley: "Physics should not make predictions for things that cannot be observed even in principle." Such things are probably mathematical artifacts, and taking them to be real can create paradoxes. In the case of cosmological inflation, the full infinity of space is unobservable. The laws of physics become self-negating when we stretch them to cover that entire volume, rather than limiting them to the patch we are in causal contact with. By confining the scope of laws to this patch—which includes parts that we might not see today but that we saw in the past or will see in the future—Bousso and his colleagues can calculate probabilities without taking the dreaded ratio of two infinities. They have found you are exceedingly unlikely to be a Boltzmann brain.[28]

Bousso's approach resembles the perspectival interpretations of quantum mechanics that I discussed in chapter 5. Because a given patch is centered on an observer, each of us sees a slightly different one. I have my reality, you have yours, and they can't be merged into a coherent God's-eye point of view. The "observers" in this approach are just possible vantage points, not necessarily active systems, let alone sentient beings. Even so, some researchers wonder whether any notion of an observer should enter into a fundamental theory at all.[29]

Besides giving up inflation or adding a measure, a third response to Boltzmann brains is to question whether these brains would be conscious. Scott Aaronson, a computer scientist and quantum-computing expert at the University of Texas, has mulled this option. He has argued that such brains would have no meaningful inner life.[30] Each of us can trace our ancestry back through our family trees, back through the history of life on Earth, back through the cosmic evolution that preceded it—all the way back to the random initial conditions at the birth of the universe. Those initial conditions eventually gave rise to you, me, and everyone else. Aaronson has suggested that the unknowable details of these initial conditions make our decision-making inherently unpredictable, fulfilling what he considers to be an essential requirement for human free will. Boltzmann brains, having no such ancestry, would then lack free will and perhaps other qualities of conscious experience. This would mean that conscious beings would be entitled to trust their observations; they're not brains in a cosmological vat.

Aaronson's reasoning is, as he was the first to admit, highly speculative. I also don't think it gets at the real nature of free will, which I will return to in chapter 7. But it demonstrates that theories of consciousness may help in exorcising Boltzmann brains from cosmology. Integrated information theory (IIT) implies much the same conclusion: a brain has to have control over its internal states to be conscious, and Boltzmann brains, being mere happenstance, lack this causal structure. In that case, there's no need to fret that you might be one.

FIRST PERSON FIRST

Müller developed his inside-out view of physics partly as a fourth response to Boltzmann brains. Unlike Aaronson, he argues that a Boltzmann brain may have some inner experience after all, and he asks us to consider what that experience would be like. To that end, he has sought a comprehensive solution to the inside/outside problem.

Müller bases his thinking on philosophical idealism: the proposition that reality is mental and the physical world is our construct. "I actually drop the idea of a fundamental world out there," he said. "I say, let's not assume there is one. Let's hope that we get it later on as a consequence or prediction of the theory. Let's start from a kind of solipsistic point of view." Most modern philosophers, not to mention physicists, detest idealism; it strikes them as mystical to assume that reality is all in our heads. Idealism also suffers from a reverse version of the hard problem of consciousness: If you assign primacy to the mind, how do you recover the physical?

Fortunately, you don't have to go along with Müller's contentious proposition. You can adopt a weaker form of his approach in which you continue to assume there is a reality independent of us and seek to describe how our brains perceive that reality. Müller's ideas then become a physicist's version of neuroscientific theories of perception such as predictive coding.

On this view, the world we create in our heads is a means to an end: to predict what we'll see next based on what we've seen before—the same problem that machine learning aims to solve. For his analysis, Müller considered an idealized machine-learning technique developed by the computer scientist Ray Solomonoff in the 1960s.[31] You take all possible algorithms that perform a calculation, check whether they reproduce your observations so far, and keep those that do. Then you see what they predict for future data and take a weighted average of all their predictions, giving the shortest program the most weight since it offers the most parsimonious explanation of the data, in accordance with Occam's razor.

Crucially, you don't take just the simplest algorithm. Simple is often right, but not always, and it's wise to keep your options open. Also, by keeping other algorithms in the mix, you obtain not just a single prediction, but a range of predictions with certain probabilities, known as algorithmic probability.

For instance, suppose you observe a string of computer bits: 11001001. There are multiple ways to interpret them. They look like a coin toss, in which case there's a fifty-fifty chance the next bit will be 0. But those bits also happen to be the start of π in binary notation—so if the data really does represent π, the next bit will definitely be 0. A coin toss is the simplest algorithm, so you tentatively assign a fifty-fifty chance to 0 or 1. Then you nudge the odds slightly in favor of 0, to account for the possibility that the data represents π after all.

Mathematicians love this procedure for its purity. Engineers hate it for the same reason: no computer could realistically run every possible algorithm. Still, our brains and AI systems do something conceptually similar. A neural network takes in data, makes predictions, and corrects itself.

Solomonoff proposed his technique as a replacement for the usual methods of physics. Instead of starting with data, formulating laws, and making predictions, you can make predictions directly from the data, cutting out the nomological middleman. Building on that ambition, Müller replaces the laws of physics with the observer's private reasoning—no universal laws required. "In physics, you normally write down an evolution equation for the whole universe," he said. "Here, there's an evolution law for that private thing alone, telling you which state changes appear with which probability." Remarkably, he thinks the internal logic of our thoughts and perceptions can account for many basic observations of the natural world.

For instance, why do the laws of physics stay the same over time? We take that for granted, but it's not a given. Müller sees this stability as a consequence of the principle of simplicity that is built into Solomonoff induction. "I can certainly make up a world which works with the same laws of physics as we have until today, and then tomorrow everything changes," he told me. "But that would be a very complicated world; I would need many bits to describe it. So that's less plausible." For those who see Müller's approach as a description of perception rather than a strong philosophical stance, the stability of nature is a rational expectation. Our best guess is that the universe will endure. It might not, but we have no good reason to expect it won't.

This principle of simplicity neatly eliminates the threat of Boltzmann brains. From a first-person viewpoint, what distinguishes being a Boltzmann brain from being a real brain is that a Boltzmann brain is transient. If you are one, the clock is ticking. At any moment, the world you know will dissolve and betray itself as a spasm of a heat-dead universe. For Müller, such a sudden shift is verboten, or at least irrational to expect. Whether the Boltzmann brains are out there is immaterial; you can feel confident that *you're* not one. "Even if cosmology would predict a huge, huge world with these Boltzmann brains out there, there would be no reason for you to believe that you disappear in the next moment," Müller said.

The same reasoning that gets rid of Boltzmann brains also works on other reality-denying scenarios. Even if you are a brain in a vat or lines of code in the Matrix, you can relax: the simulation is so true to life that you can treat it as real. "I would say there is no ontological difference between us being a brain in the vat and making exactly the same observations as we do now, and us being an actual thing in the world and making the same observations," Müller said. His point is that, if you take the first-person view as primary, two experiences that seem the same *are* the same, period. It makes no sense to deem one real and the other illusory. In this, Müller agrees with Rovelli and others I mention in chapter 5 who take relations to be the fundamental ingredients of nature. For them, there's no meaningful difference between a thing and a hallucination that expresses all the same relations.[32]

To be sure, Solomonoff induction can't explain everything. The detailed workings of the world aren't the products of generic reasoning and could well have been different. "It's not a theory of everything," Müller said. "It's not possible to predict what the laws of nature will be; it's contingent what they are."

One of the most interesting aspects of Müller's approach is that it answers the question I posed in chapter 5: Why do observers ever agree on anything? If each of us has our own reality, why don't I notice that my world is different from yours? "Typically, in physics, we think, of course, How could it be different, because there's just one world out there?" Müller said. "But [in my view] it can be different. So you first have to prove as a theorem that in most cases it's not different." He has shown that if observers are using Solomonoff induction and working off the same data, they will reach the same conclusions. Rational thought—understood as the simplest mechanism that produces accurate predictions—overcomes our differences of perspective and lets us reach consensus. "We are hallucinating in a consistent way," Müller said, "and that's what we call the world around us."

From this excursion into cosmology and its puzzles, we take away a basic lesson: when physicists don't take the act of observation into account, their theories are sterile; they are disconnected from the very reason we do science, which is to explain what we see. Sometimes we find that our theories are at odds with our observations, not because they are wrong, but because, lacking the human element, they are incomplete. This same issue comes up in the domain of causation.

7

LEVELS OF REALITY

IF THERE'S ONE THING YOU expect from physics, it's an explanation of what causes what. Physicists wouldn't be physicists
if they didn't have the intuition that what happens, happens
for a reason. They are storytellers, spinning out a narrative
for what makes planets orbit, bombs explode, and electricity
spark. They speak of forces, interactions, reactions, excita-
tions, changes of state—all causal notions. When I was in
college, professors taught us to use principles of causality as
a reality check on our mathematical derivations. If my solution to a homework problem implied that something stood
outside the causal order, unable to affect other things or be
affected by them, or predicted a causal anomaly such as an
effect happening before its cause, I crumpled up the paper
and tried again.

So I was faintly disturbed when, while working on my first
book, I learned that many physicists think causation is an illusion.
"Physicists rarely think that the past 'causes' the future," Carlo
Rovelli told me in 2008. "There are regularities expressed by laws.
That's it." For years I kept this issue on my to-do list, and during the
pandemic Holly Andersen, a physicist turned philosopher at Simon
Fraser University near Vancouver, filled me in on the backstory to

Rovelli's remark. She recounted how misgivings about causation multiplied over the course of the Scientific Revolution and culminated in a famous paper by the English philosopher Bertrand Russell in 1912. "Russell came along during his 'I'm going to piss everybody off' phase," she told me. "He's this young guy who says, 'Well, this principle of causality—we don't even need this anymore.'"

All that the laws of physics do, Russell argued, is describe patterns: relations among mathematical variables. Consider the ideal-gas law. It tells us that if the pressure is 100,000 pascals and the temperature is 0 degrees Celsius, then if the pressure drops to 37,000 pascals, the temperature will drop to −172 degrees. "You can calculate what the values of these things are going to be, but there isn't, as it were, a privileged one—which one is the control variable for the other ones," Andersen said. "They're all in a symmetric functional relationship." Because the variables stand on equal ground, one doesn't compel a change in another. The law is also agnostic about why the variables change. The gas pressure might drop because a human ran a vacuum pump or because a human used a refrigerator to cool the gas, which in turn reduced the pressure; the outcomes are entirely equivalent. For Russell, the adage that "correlation does not imply causation" was too tame. For him, there is never any causation, only correlation.

Furthermore, Russell noted that whereas causation is an asymmetric notion, the laws of physics are symmetric within time—events can be run forward or backward, swapping cause and effect, rendering the distinction meaningless. How can you claim you sank the eight ball if you could equally well say that the eight ball spontaneously leapt out of the pocket and struck the cue ball, which rolled across the pool table and hit your cue stick? We never see these events in this reversed order, of course, and one of the top puzzles in modern science is why not. But the answer does not lie in the laws of physics per se.

As if that were not enough to doom causation, laws in physics are global. They say that what happens on Earth today is the product of the entire state of the solar system yesterday, which undermines the concept of an isolatable cause. If you sink the eight ball, you need to give due credit to your friends who challenged you to a game, to the pool-table manufacturer, to the sun for not exploding while you were making the shot—where does it end? Depending on how far back you trace the reasons, everything in the observable universe played some part. And if everything causes everything, then nothing causes anything.

Russell closed his case with a devastating putdown: "The law of causality, I believe, like much that passes muster among philosophers, is a relic of a bygone age, surviving, like the monarchy, only because it is erroneously supposed to do no harm."[1] Few dared disagree.

Over the following decades, quantum physics deepened the conundrum. (It often does that.) If a photon strikes a half-silvered mirror, your odds of seeing it reflect off or pass through are fifty-fifty. There's no reason you'll see one outcome rather than the other. The outcome is indeterministic; it has no cause. That is true across all interpretations of quantum theory, although they disagree on precisely what kind of indeterminism is operating. This causelessness in turn creates other puzzles. For example, what makes quantum entanglement so mysterious is that the fates of distant particles are linked even though the theory specifies no mechanism to link them.

So physicists face a paradox of causation. They can't live with it or without it.

REALITY IS A LAYER CAKE

Whatever physicists may think, the rest of us sure can't live without causation. Our every action has consequences, or so my mother told me. If physicists deny causation at a fundamental level, it behooves them to explain why she put me in time-out after that episode when I stole a penny from her coin purse. More generally, they must explain why we observe a cascade of causes and effects on every scale. At the cellular level, a buildup of ions causes a neuron to fire. At the human level, seeing a cookie causes you to crave it. At the economic level, increasing demand for cookies causes their price to rise.

That the same phenomenon—a hankering for cookies—can be viewed on so many levels is one of the most incredible facts about nature. In principle, nature might have had just one level: the particle level. If so, we could scarcely exist, let alone comprehend anything. One particle is easy to deal with. Two particles make things interesting. Three—even just three—can pose an intractable problem. Dozens, forget about it. Trillions, oh come on. Yet it turns out that trillions of particles appear to behave collectively according to simple rules, allowing us to get our heads around behavior that would otherwise be beyond us. Because of the group dynamics of particles, new, higher levels of description emerge that clearly depend on the lower levels, yet stand independent of them. Psychology presumes a brain but depends very little on the details of neurophysiology, and economics couldn't care less what a cookie is as long as people are willing to slave away in a cubicle to buy one. The autonomy of levels is what justifies calling them levels.

Some researchers take the reductionist position that the higher levels are really just an illusion. If you eat a cookie, what is really going on is that a ginormous number of particles move in a mostly coordinated way. There is no such thing as "you" or "cookie," but we create these mental categories because our puny brains couldn't keep track of all those particles directly. "They are rather like the clouds that appear in the shape of animals," the philosopher William Seager wrote.[2]

Others beg to differ. Taking what is known as an emergentist view, they think that "real" should not be equated with "root" and that genuinely new properties emerge at higher levels of organization. The success of chemistry, biology, psychology—any science other than fundamental physics—would be a miracle unless there were some real structure at those levels. What is more, with every breath and bite, our bodies swap some particles for others. Most of your body is less than ten years old, no matter your age.[3] That our bodies endure despite this ceaseless turnover indicates they are something above and beyond particles.

These opposing attitudes toward the reality of higher-level structures and causation are often identified with physicists and biologists, respectively, but the line is drawn more by individual temperament than by discipline. Plenty of physicists are emergentists, and plenty of biologists are hard-core reductionists.[4]

It's striking that the debate is conducted largely in a conceptual vacuum. Whatever their intuitions about levels of causation, scientists are seldom entirely clear on what causation actually is. This haze is thickest in debates over free will. People make sweeping pronouncements about human agency, or lack thereof, without explaining what they are really talking about. Those who think we have free will assume an emergent concept of causation that they need to articulate better. Those who insist we have no free will argue that causation occurs only at the base level of nature, but if Russell is right, that's exactly the wrong place to look for causes and effects; the fundamental laws deny causation makes sense at all: "If you don't have a good measure or concept of causation, then how are you sure about your claims that the microscale is necessarily doing all the causal work?" asked the neuroscientist Erik Hoel.

MORE IS DIFFERENT

I was introduced to Hoel in 2016 after a panel discussion on the science of consciousness. To my surprise, the first thing he wanted to talk to me about was literary agents. He had recently

moved to New York to start a neuroscience postdoc and was working on a noir novel about a young scientist who moves to New York
to start a neuroscience postdoc. I put him in touch with a novelist
friend, and his genre-crossing novel came out in 2021. He later left
academia in order to pursue writing full-time. In return for making
the connection, Hoel introduced me to new ways of thinking about
causation.

Hoel had been a grad student of Giulio Tononi, the father of
IIT. Contrary to Russell, Tononi thinks causation is essential—in
fact, it is the philosophical foundation of his theory, the ultimate
justification for his conception of consciousness as the integration of
information.

Your mind exists—that is, as Descartes famously observed, the
one thing you can be sure of. To have a thought implies a thinker.
But what does it mean to exist? In answering this question, Tononi
has been strongly influenced by physics. Physicists are inclined to
think that nothing in existence is inert; if something neither affects
nor is affected by other things, it might as well not exist. Whenever physicists have come across an apparent exception to this rule,
they've discovered that the thing indeed doesn't exist—or else that
it isn't so passive after all. "No physicist would hope to claim that
a fundamental particle exists, that has been discovered, if it were
not the case that you can manipulate it somehow and observe it
somehow," Tononi said. "So, if you cannot have causes and produce
effects, well, it's like talking about angels."

Einstein provided the classic example of the connection between
existence and causal power. Physicists used to regard space and time
as a fixed backdrop to the universe, but Einstein realized how strange
it was to place space and time outside the causal flow of the world;
that made them vaguely supernatural. In 1922 he wrote, "It is contrary to the mode of thinking in science to conceive of a thing (the
space-time continuum) which acts itself, but which cannot be acted
upon."[5] Einstein challenged the inertness of the continuum with his

general theory of relativity, which showed that space and time are warped by energy, and are thus as physical as anything else.

Around the same time, the British philosopher Samuel Alexander argued that the mind, too, must have a causal role, and that existence in general entails causal power.[6] Philosophers today refer to this idea as Alexander's dictum or, based on similar ideas articulated by Plato, the Eleatic principle.[7] The idea is that if you observe that something exists, you know it must do something causally. Ergo, your mind must be a player and not just an onlooker in the drama of life. In the early 1960s the physicist Eugene Wigner, whom we met in chapters 4 and 5, invoked the same principle when he argued that it made sense for consciousness to play a role in quantum mechanics: "We do not know of any phenomenon in which one object is influenced by another without exerting an influence thereupon."[8] To be sure, the principle wasn't central to his argument, and he later questioned whether consciousness did have such a role.

For Tononi, the mind exists inasmuch as it can make things happen—if not in the external world, then at least in its private mental domain. For him, the brain must have control over its own mental states to be conscious. It must drive its own sequence of thoughts; you can't be a passive spectator in the theater of your mind.

One of Tononi's colleagues and close collaborators, the computational neuroscientist Larissa Albantakis, who helped to develop IIT and work through its implications, told me that the moment our mental states slip out of our control, they cease to be experiences. "If there is an evil neuroscientist that forces your neurons to be in particular states, they actually cannot contribute to consciousness anymore, according to IIT," she said. "This is precisely what gets evaluated by IIT: that the system itself has a potential to influence its own states. If those states get determined from outside, the system cannot be conscious."

The brain does take in information from outside: light striking the retina, sound reverberating in the ear. When Tononi and

Albantakis's colleagues stimulated my brain with their magnetic coil back in chapter 3, they bypassed the normal channels of information flow and manipulated my neurons directly. But despite such external influences, Albantakis said, the brain remains conscious because it never fully cedes control; our mental life is informed by, but not pre-scribed by, what happens around us. "Things outside trigger things inside," she said.

According to IIT, because the mind can act and be acted upon, it exists in exactly the same way any other material object does. At first glance, the human mind seems to fall into an entirely differ-ent category from, say, a table. Its properties, such as the capacity to experience red or love, seem ineffable compared to the table's heft and physicality. Yet both the table and the mind are defined by their structures. What makes a table a table? It isn't the individual atoms, because most of the properties of the table are not those of the atoms. Its solidity is a collective property—a product of how the atoms are locked together like Legos. The very same atoms, under different conditions, might crumple onto the floor or go up in smoke. By vir-tue of the table's structure, we are entitled to think of it as something above and beyond the atoms in it. The same goes for the mind, ac-cording to IIT: the theory identifies the mind with the structure that a network can exhibit—namely, its information integration—giving it properties that neurons or other units lack in isolation. "Once you plug them together in a certain way, there is something new under the sun," Tononi said.

To summarize, the mind exists, so it must have causal power, including control over its own stream of consciousness. It cannot be a purely reactive system, enslaved to the outside world; it must have some meaningful inner dynamics, which IIT quantifies in terms of information integration. And on account of this causal power, the mind stands as an equal with the usual objects of physics. In this way, IIT connects philosophizing about existence and causation to empirical predictions for neuroscience and AI.

Because causation is so essential to IIT, it forces the issue of figuring out what causation really is and which level or levels it operates on. And in articulating a concept of causation for the brain, IIT
may help to unpick some of the mysteries of causation in physics,
too. "This is where I think it's not just a theory of consciousness,"
Hoel said.

In seeking to extend their ideas to physics, he and his colleagues
again follow Samuel Alexander. Alexander belonged to a loose
school of philosophers in the late nineteenth and early twentieth
centuries, the British Emergentists, who saw the mind as a case study
for the emergence of higher-level properties in material systems.[9]
In fact, they were the first to use the word "emergence" to refer to
principles of collective organization.[10]

Physicists at the time were oblivious to these philosophical over-
tures. They became interested in emergence much later, prompted
by a 1972 essay titled "More Is Different," by the theoretical physicist
Philip Anderson.[11] Anderson was driven largely by academic politics within physics and, in particular, a desire to show that his own
specialty—the study of solids, liquids, and other materials (otherwise
known as condensed matter physics)—was every bit as foundational
as particle physics.[12] He never used the word "emergence" in that
paper, however, and later admitted to having been unaware of all
the good work done by earlier scholars.[13] Anderson was invited to a
neuroscience meeting in 1977, and the disciplinary walls have been
crumbling ever since.

RETHINKING CAUSATION

Continuing in this tradition of crossing fields, Tononi, Hoel,
and their colleagues have embraced an influential theory of causation
that has swept through statistics, philosophy, and computer science
in the past fifteen years. It is known, in various guises, as interven-

tionism, causal perspectivalism, or agency theory. It accepts Russell's point that cause and effect are not fundamental categories, but considers them real all the same.[14] The theory holds that causation arises at the level of human and other actors, and for a simple reason: it depends on those actors.

The reason we care about causation is the need to get things done. People have to know which button to push or lever to pull. "What we're looking for isn't just lawlike connections," Holly Andersen, the philosopher at Simon Fraser, told me. "What we want is an effective means of controlling." The fundamental laws of physics, on their own, don't tell you that. They predict what a system *will* do given its present state, but not what a system *could* do—what would happen if you manipulated it. That depends not only on the laws, but also on the system's internal structure: how its parts mesh with one another.[15]

You take one of these parts as your lever and force it to do something it might not have done in the natural course of events. The other parts, being mutually correlated, should change in response. "If we change the light switch from on to off and we wiggle it up and down, separately from what it would have been at before, then we can see what's causally downstream of it," Andersen said. The asymmetries we associate with causation are our own doing. What is "cause" and what is "effect" hinge on how you decide to act on the system. You can choose to manipulate another part instead or even redefine what the parts are, and that will change the causal categories.

Even to speak of causation requires that you divide the world into "you," "the system," and "all the rest," and treat your own actions as freely taken. Someone looking on may see things differently. They might consider both you and the system to be part of some larger system, and might find, for example, that the system caused you to pull the lever rather than the other way around. So causation is a first-person phenomenon—a product of your vantage point—and the apparent paradoxes of causation are a subset of the inside/

outside problem. For proponents, this interventionist view is another example of Kant's insight, introduced in chapter 1, that many basic features of the world are as much about us as about the world.[16] For skeptics, though, it seems unduly anthropocentric.[17] To avoid this latter charge, the "you" who intervenes in a system need not be a conscious agent, but simply any part of the universe that stands outside the system.

One point in favor of the interventionist view is that it comes with a deep toolbox of mathematical techniques. No longer stuck wagging their fingers that correlation does not imply causation, statisticians can map out webs of cause and effect using diagrams with lots of arrows indicating which variables affect which others. Guided by these diagrams, they turn the original symmetrical, global, causeless equations into asymmetric, local, cause-effect relations. These methods go back to the '20s,[18] but came into their own as technology made them easier to implement. "Between 2000 and 2005, [the interventionist view] just rapidly coalesced, and I think it overlapped, not accidentally, with dramatic increases in data-analysis possibilities on computers," Andersen said. "You used to have to argue for the idea that we should talk about causation at all, that there was such a thing as causation, that it wasn't just talking about mechanisms, that you could use this idea of intervention. The talks that I had to give used to have to justify a lot more features of causation than they do now. Now I don't even have to tell people about interventionism."

Judea Pearl, a computer scientist at UCLA and the doyen of causation studies, cited the example of determining a causal link between cigarette smoking and cancer. This was a dark chapter in statistics. In the '50s some leading statisticians said you couldn't tell whether smoking causes cancer, cancer causes smoking, or some third factor causes both; all the data showed was an association between the two.[19] (They also denied any causal link between their skepticism and tobacco-industry funding.) Perhaps they were just being prudent—we humans do have a tendency to interpret coincidence as evidence of causation. But when lung-cancer rates are skyrocketing, you'd hope that statisticians could do more than throw up their hands. Pearl showed that, by considering additional variables that are part of a possible causal mechanism, statisticians can nail down the direction of the effect. For instance, if you conjecture that smoking causes cancer by depositing tar in the lungs, you can go in and measure those tar deposits by, for example, testing lung function.[20] In the context of this hypothesized mechanism, a correlation between tar deposits and cancer incidence *would* imply causation. In effect, you can find naturally occurring controlled experiments hidden within passive observations.

Pearl has offered other examples of how the interventionist view avoids statistical paralysis. People often think of causation in binary terms: either something is the cause or it isn't. This often comes up when science weighs in on public policy. Did climate change cause a continental heat wave, yes or no? When we instead think of causation in terms of interlocking variables, an event can have multiple causes. Natural weather variability and human activity are *both* to blame for heat waves. Many extreme events wouldn't have happened if not for greenhouse gas emissions.[21] We don't need to get caught up in trying to find "the" cause before we act.

The interventionist view also organically handles different scales in nature. Any time you exert control over a system, you are acting causally; the control does not need to happen at a particular scale. Christian List, a philosopher currently at Ludwig Maximilian University who studies causation, gave the example of human actions. If you ask a taxi driver to take you to Paddington Station, she will (hopefully) take you there. What caused her to do that? Reductionists would say it was the pattern of her brain activity or, if they're really hard-core, the motions of electrons, strings, or whatever the fundamental building blocks of reality are. Most of us would say simply that the driver heard your request and acted on it.

Both are right, although the higher-level, psychological explanation makes the connection between cause and effect clearer. For one thing, it explains variations more readily. If you ask for St. Pancras Station instead or make the request of a different driver, the specifics of brain activity will differ, yet the psychology will be basically the same. All you have to do is say the word, and off you go. As hard as it can be to persuade people to do something, it's still easier than attempting to bypass their senses and alter their brain activity directly. You would have to zap billions of neurons, each in just the right way. "If I want you to do something for me, I don't do this by manipulating your brain state," List said. "Of course, it's totally unethical to do this, but secondly, it would also be totally infeasible. The most systematic and reliable way in which I can get you to do something is by, for instance, asking you to do it."

FOUR QUARTERS OR TEN DIMES

Neural networks make excellent sandboxes for studying causation on multiple scales. Using them, you can test claims that are otherwise beyond you. For instance, it would be hopelessly ambitious to show directly that particle physics gives rise to human psychology—you'd have to bootstrap all of chemistry and biology first. The gulf is just too wide. We have to take it on faith that chemistry and biology do derive from physics; this has never been strictly proved. But neural networks embody in microcosm the same principles of collective organization that occur in nature. "What's cool about machine learning, for me, is that the distance between the microscopic model and the output is much, much smaller," said the physicist and neural network researcher Dan Roberts. "There's a bunch of steps, but it seems much more approachable than going through chemistry and biology in the middle."

Hoel runs with this idea. He considers some very simple cases to explicate the sometimes mysterious process of emergence. He starts with a network like the others we've been talking about and takes it as the fundamental description of a system. Then he studies how the units turn on and off in response to one another, seeing whether the units might be grouped together in ways that create simpler but equivalent networks. He borrows from physicists' standard method for navigating multiple levels of description, known as renormalization, while adding distinctive concepts.[22]

Hoel gave the example of a miniature network: basically, a pair of lightbulbs screwed into a two-socket table lamp. The network has four possible states: both on, both off, one on and the other off, and vice versa. If the bulbs have separate switches, they don't really form a network; they are just two independent lights. But often these lamps are wired together so that, by turning a single switch, you cycle through the four states. Then you have a true network that you can describe either at the level of the individual bulbs or at the level of their combined effect. Hoel zeroes in on three ways that the two levels might be distinct.

First, the higher level can dispense with irrelevant details. Suppose, for example, the bulbs are wired so that only one is on at any given time. This little network always provides one bulb's worth of illumination. If all you care about is having enough light to read by, it doesn't matter which bulb is on. The irrelevance of details is known to philosophers as multiple realizability and to physicists as universality or substrate-independence. One example is that a dollar bill, four quarters, and eight dimes and four nickels are multiple ways to realize the same amount of money. A dollar is a dollar, whether made of copper, paper, or digital bits—its value is independent of its material incarnation.

Such compositional flexibility is ubiquitous in nature. A glass of water looks placid, yet we know it is a vast, heaving ocean of molecules—more of them than all the humans who have ever lived, colliding trillions of times per second. Their complexity is hidden from us not just because molecules are small, but because their zillions of arrangements, from here at our scale, all appear basically the same. If you move one molecule a bit to the left or swap it with another, the molecule might notice, but no one peering at the glass from the outside will be any the wiser.

By neglecting these molecular machinations, you greatly simplify your description. Simpler means more deterministic, which means tighter causal control. If you shake, stir, or squeeze the water in bulk, you can reliably predict the outcome, which is very hard to do at the level of H_2O molecules if you act on those molecules one by one. To be sure, any high-level description has its limits; if you heat the water to a boil, you'll have to switch to a new high-level description. But over its range of validity, each of these descriptions illuminates the essential physics that would otherwise get lost in the molecular weeds.

A second way that wholes can be more than the sum of their parts is redundancy. A system can start off looking very complicated, but settle into one of only a few states. The other states never recur, and you gain in explanatory clarity by neglecting them. In the two-bulb network example, suppose one of the bulbs blows out as soon as you turn it on. From then on, you can forget about it and treat the system as a single bulb. This kind of attractor dynamics is common. We saw it in chapter 2 with the Hopfield network, which has multiple stable patterns of neural activity and will transition to one of them. These patterns are usually all you need to describe the system.

Finally, the higher level can take advantage of modularity. When a group of components performs some specialized function, you can treat it as a single unit and forget its inner workings, much as you can regard a living being as a collection of cells or computer software as a series of standardized subroutines. Hoel and his colleagues thus invite physicists to think like biologists or software engineers. George Ellis, a theoretical physicist and mathematician at the University of Cape Town, told me that this insight is an important addition to standard renormalization theory: "They are taking seriously the modular hierarchical nature of complex structures, which is the key to complexity."

Whereas stripping out irrelevant or redundant details makes a system more deterministic and predictable, modularity can make it less so, because sometimes the "function" of a module is to create noise. For instance, a coin toss is completely deterministic if you think in terms of the basic physics: the air currents in the room, the precise flick of your fingers, and so on. But those details are hidden from you, so they don't help you to predict the outcome. A higher-level description treats the coin toss as truly random. (This is separate from any randomness that quantum effects might add.) Life is filled with situations that are so complex that they are effectively random, and you save yourself a lot of frustration if you treat them as such from the get-go, rather than pretend you have control.

Having streamlined your description of a system, you can repeat the process, looking for additional structure and moving to an even higher scale. Hoel showed that you gain explanatory traction by going to a higher-level description. Appealingly for physicists, he puts a number to the gain. For the two-bulb network, if the bulbs cycle through all four of their possible states in succession, you have perfect knowledge of what they will do when you turn the lamp's switch knob. For two bulbs, that's 2 computer bits of information. But other situations are less certain. Suppose a wire is loose, so that if both bulbs are off, they stay off, but otherwise they flicker randomly. Working through the math, knowing the system's current state gives you only 0.81 bit of information about its future. The connection between cause and effect is weakened.

To restore predictability, you collapse the three randomly cycling states into one. The new network is smaller, just two states—"off" and "flickering"—or a single computer bit. But now it is fully deterministic. So knowing its present state gives you 1 bit of information about the successor state, for a gain of 0.19 bit from the original description. "The higher scale is not just a compressed description,"

Hoel said. "Rather, it's that by getting rid of noise, either in the case of increasing determinism or by reducing redundancy, you get a more informative description." Something similar happens by the zillions with molecular motions in water.

HOW IS CAUSATION LIKE TEXT-MESSAGING?

This IIT-based approach to causation challenges many intuitions that physicists have about emergence. First, it dissolves the

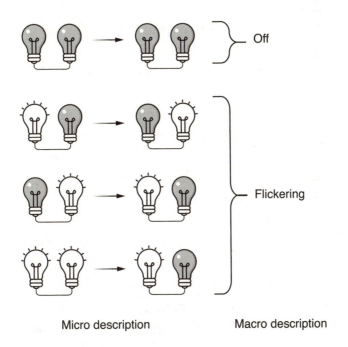

Micro description   Macro description

7.1. CAUSAL EMERGENCE. Consider a rudimentary network consisting of a pair of lightbulbs controlled by a glitchy switch. If both are off, they stay off, but if one is on, they flicker randomly while never going entirely dark. So the network really has just two states: "permanently off" and "perpetually flickering." This setup illustrates how neglecting randomness can reveal a system's essential dynamics. This is a simple principle of emergence that also operates in much more complex systems.

intuition that causation must occur either at the base level or at the high level. It can happen at both. Hoel's mathematical method apportions causation among multiple scales. In the flickering two-bulb network, you could say very loosely that 81 percent of the causal oomph of the network lies at its base level and 19 percent at the higher level.

A second common intuition is that higher levels must contain less information since, by definition, they gloss over details. In Hoel's analysis, a higher level does lose information by being simpler, but it also gains information by being truer to the network's structure. Less is more. Joseph Halpern of Cornell University, a computer scientist who worked with Pearl to develop the interventionist theory of causation, said: "Hoel is essentially pointing out that a 'small' model may be more than just an approximation of a larger model. It may actually in some sense of the word have more information than the larger model."

To buttress this point, Hoel has drawn on theorems from an unexpected quarter: communications engineering. Signaling is a type of causation. You tap a key on your phone and cause a letter to appear on your friend's screen. Making that happen reliably requires sophistication: our messages punch through electrical interference only because our phones and devices encode data in a form that resists degradation. "Code" just means a way to represent information. Morse code, for instance, translates letters and numbers to dits and dahs and then to electrical pulses. It takes advantage of the structure of the English language, encoding the most common letters, E and T, with the shortest sequences to speed up the transmission. Predictive coding is a more sophisticated version of that. Other types of codes compensate for errors in transmission, based not on the structure of the data, but on the characteristics of the medium.

Codes create levels of abstraction. We don't have to transmit our messages using raw physical states, but can cleverly combine those states to squeeze the maximum performance out of a system.

Hoel showed that levels of causation are entirely analogous. Higher levels of causation scrape away the noise in a system—the irrelevant details—to let the essential dynamics shine through. Using them, you maximize the control you exert. "Higher scales offer error correction by acting in a similar manner to codes, which means there is room for the higher scales to do extra work and be more informative," Hoel said.

A real-world example that closely resembles Hoel's simple two-bulb network is computer flash memory.[23] At the microscopic level, it looks like an egg carton, with rows and rows of units. Each unit has four different voltage levels and, in principle, could hold two computer bits. But only one of these voltage levels is reliable, while the other three tend to cycle among themselves. So instead of trying to cram in a pair of bits, engineers store just one in each unit, encoding 0 as the reliable voltage and 1 as one of three flaky ones. Such a code halves the storage capacity, but what good is storage capacity if you lose your data? By analogy, a higher-level causal description may gloss over details, but those details often don't really matter.

A third common intuition that Hoel, List, and others have revisited is that determinism is an either/or situation. People often say the world has to be either intrinsically random or regular and predictable. Much of the debate over quantum physics hinges on which it is. But in fact the world doesn't have to be one or the other. It can be both. "The world could be deterministic at one level of description and it could also simultaneously be indeterministic at another level of description," List said. This is because each level reworks the laws of nature. The laws governing liquid water, for example, are a product not just of the laws governing the individual H_2O molecules, but also of the way those molecules are arranged. Even if the molecular laws are fully regular and predictable, higher levels can become randomized. Within raging river rapids, the molecules may be entirely orderly. Conversely, a smoothly flowing fluid may belie erratically waggling molecules.

This concept of the level dependence of determinism opens up a remarkable possibility. Might there be no fundamental laws of physics at all? In Hoel's model, a deterministic higher-level description can emerge from a base level that is thoroughly anarchic. "Effective information can be unbounded at the macroscale while approaching the limit of zero at the microscale," Hoel said. His work thus lends credence to the physicist John Wheeler's idea of "law without law." Wheeler speculated that chaotic microscopic events "flaunting their freedom from formula" are nonetheless collectively law-abiding, with "billions upon billions of such acts giving rise, via an overpowering statistics, to the regularities of physical law."[24] If so, physicists could dig to the foundations and find that reality is built on quicksand.

As helpful as Hoel's scheme is in clarifying the concepts of emergence, it doesn't get us very far in figuring out how emergence works in most real-world situations. Hoel considers very simple networks and even then has to resort to a brute-force computer analysis to identify the multiple scales on which they operate. Finding the structure within a system is fundamentally hard, because there are so many possibilities to consider.

Irina Higgins, a neuroscientist and an AI researcher at Google DeepMind in London, sees connections between Hoel's work and her own research, which aims to help artificial neural networks pick out the right structure in images. If you train a network to identify cats, it will dutifully spit out a label for any animal you show it, but that doesn't mean it has identified catlike structures within the images. It might instead be creating bizarre combinations of pixels that happen to be correlated with the type of pet in those images, rather than conceiving of cats as creatures with tails, fur, and whiskers. Higgins is able to force a network to create realistic representations of cats. Her techniques, like Hoel's, work by eliminating redundancy, on the assumption that a parsimonious description is truer to reality.[25]

But these techniques don't yet work for multilayered images. "I am not aware of any model that can do it properly right now," she told me in 2021. She gave me the example of an image showing sheep in a field surrounded by forest. "Do you represent each sheep or a flock as a whole? Do you represent the background as a single item, or do you split it into field plus forest plus sky, or do you represent it at the level of individual trees or even blades of grass?" she said. The machine has no reason to parse the scene one way or the other. Our brains would handle the task almost effortlessly, but even they build in a lot of presuppositions about what the structure is likely to be. For the same reason, it is inherently difficult to separate causation by layer. We take it for granted there is one way to make the world, when really there are countless ways.

IS IT POSSIBLE TO SAY ANYTHING NEW ABOUT FREE WILL?

A more sophisticated understanding of emergence could also help to unstick debates over free will. Traditionally the domain of philosophy, free will is a concern of physics, too. If we are to achieve a full understanding of causation, we can hardly leave out the most intricate causal actors known to science: ourselves.

Free will is the rare philosophical concept that is useful as well as fascinating. For example, our justice system and democratic processes are built on assumptions about individual volition. And debating whether we have it used to be almost as good a way to pass an evening with friends as playing Cards Against Humanity. But free will has lost some of its fun for me; these debates can bring out the worst in people. Everyone seems so sure of themselves.

But if everyone just chills, they will see that progress is possible. Consider how the debate has evolved historically. The question of whether we are the authors of our choices or merely cogs in a clockwork universe—or whether those two options are truly in opposition—used to hinge on determinism: If everything that happens, happens for a reason, then whether I'll choose coffee or tea tomorrow morning is preordained. Eons before I or coffee or tea or Earth existed, the atoms that filled the early universe were subtly imprinted with the imperative, Get this man some coffee. But if people debating free will have come to agree on anything, it's that determinism is a red herring. The laws of physics may well be indeterministic, in which case my choice of tea or coffee is due to a random atomic swerve. From my point of view, that's no different from deterministic preordination. The choice is still being made for me.

Another point of general agreement on free will is that humans aren't exempt from the laws of physics. The eminent philosopher who taught my college metaphysics class in the '80s tried to convince us that we stand outside physics—that we have free will because human agency is, like God, an unmoved mover.[26] Few think that anymore. Most accept that our decisions don't break physics. They aren't bolts out of the blue. They have antecedents.

Today the debate has shifted to the nature of causation. If causation lies entirely at the fundamental physical level, then we're just puppets, or not even that—just big puppet-shaped blobs of atoms. But if Russell's critique of causation is right, physics at its roots has no directedness or sense of compulsion; the category of cause just doesn't apply there. Some physicists do think that causation is fundamental after all.[27] But even if they're right, the causes and effects that are relevant to free will arise at a higher level of description, just as human beings (as opposed to blobs of atoms) arise at a higher level of description. We can't talk about your making a decision until we can talk about *you*. For Hoel, List, and others, higher levels do have causal power, and so, potentially, do you.

So the real question is whether being part of the causal flow of the universe makes us unfree. If your definition of "free" is "physically uncaused," then yes, it does. Some people do adopt that definition. But I find it a strange position to take: the whole point of a free choice is that it is *caused*—by you. It flows from your desires and deliberations or maybe your impulsive choices. What greater freedom could you have? "You *want* your brain to make you do it," the Tufts University philosopher Daniel Dennett told me. Not only is being part of the causal flow compatible with freedom, it is also necessary for freedom.

There is irony here. One reason that free-will detractors get so hot under the collar, I think, is that they see free will as a pre-scientific holdover—a vestige of a mystical worldview that not only separates mind from matter, but also places minds beyond scientific understanding. But they, I find, are the ones who adopt such a dualistic view. They assume that free will would require the stirrings of an incorporeal soul unconstrained by physics, so if you rule out souls, you rule out free will, too. But all you really rule out is that particular conception of free will. If you instead assume that consciousness is ultimately physical (perhaps in a broadened conception of physics), our desires and deliberations are the outcomes of physical processes; yet as long as we act on the basis of those desires and deliberations, we are acting freely.

That leaves the sticky issue of whether desires and deliberations that are caused by prior events can really be ours. One way to think about this is to consider what Dennett calls the Cartesian theater.[28] You might think that consciousness is like a movie that you watch inside your head. But who would be the viewer? That inner self would have to have conscious experiences in order to perceive the movie, requiring a theater within the theater, and so on. A dualist would break the infinite regress by imagining a soul implanted in the body. For a physicalist, though, it would be better to say that you can't separate your self from the experiences that you have; the two come as a package.

Applying the same reasoning to free will, for a physicalist it makes no sense to think of yourself as a puppet who is created and then manipulated. You do not precede your thoughts, decisions, and actions. Rather, you *are* your thoughts, decisions, and actions. The same life experiences that lead you to act in a certain way also create your self. So again, being part of the causal flow enables rather than crushes free will. Our decisions are the outcome of previous events, but they are a unique confluence of those events; no two people share precisely the same life history. These differences make us individuals and justify calling your acts *yours*.

Free will is another instance of the inside/outside problem. Physics traditionally provides us an outside, objective view, but for a full explanation of reality, you have to connect that perspective to the experience of an embedded agent. From the outside, human behavior may be fully predetermined. When my daughter was a toddler, I could predict with near certainty what she'd do when we put salad on her plate: she'd spurn it. But her predictable refusal did not deprive her of agency; to the contrary, it meant she was exerting her agency. Until we make our choices and act on them, the outcome is open from the inside perspective. Jenann Ismael, a Johns Hopkins University philosopher of physics who has analyzed free will, quantum physics, and much besides, argues that freedom is performative: to do is to have. "Judgments, choices, decisions: these have the status for the chooser that, for example, the verdict rendered by a jury has for the jury," she told me. "They are made true in the act of affirming them."

FREE WILL, MEASURED IN BITS

Doubtless the free-will debate will go on. Not everyone is convinced that making choices according to our desires counts as free will in a metaphysical sense, but these ideas spin off lots of questions that scientists might make headway on. For instance, scientists can try to measure free will.

Like causation, free will is not binary, but comes in degrees. People can be compelled or constrained in their decisions by varying amounts. Tononi, Albantakis, Mélanie Boly, and their colleagues have suggested using IIT to apportion causation among different levels: a human or other agent accounts for some of causation, and that is the agent's free will.[29] Alternatively, Ian Durham, a physicist at Saint Anselm College in New Hampshire, has quantified free will in terms of available choices: the more constrained we are, the less freedom we can assert.[30] On a more practical level, cognitive scientists have proposed a "free-will index": They would administer a battery of psychological tests to assess people's capacity to deliberate and act. Some people who commit crimes might have acted completely freely, and the justice system might as well throw the book at them. But others might not have acted completely freely and might benefit from rehabilitation to help them make better decisions.[31]

Quantum physicists have their own reasons to assess free will. For them, the issue is how to interpret their famously bizarre experimental results, such as those I described in chapters 4 and 5. These experiments, like those in any science, implicitly assume free will, since they involve choices about how to conduct the experiment or assign trials to a control group. In this context, "free" doesn't have any metaphysical intimations; it just means unbiased. Some physicists worry we do not have free will in this sense—the view known as superdeterminism. One way to check is by running the experiments and asking how much of a constraint on our freedom would be needed to explain away the results.[32] Even a very slight constraint would be enough, researchers have found. "They transformed the qualitative question into a quantitative one," said the quantum physicist Sandu Popescu of the University of Bristol.

Physicists are also asking what kind of physical system is capable of agency—of being able to act on its own accord, possessing free will or at least the illusion of it. Simply being a link in a causal chain is clearly not enough, or else everything in the universe would be an agent. The special sauce is the sophistication of our minds. Most physical systems are reactive, meaning that they respond only to their immediate circumstances and affect only their immediate surroundings. Causation in these systems is straightforward; effects are proximate and proportional to causes. But intelligent beings and artificial neural networks create twisty paths between cause and effect.[33] We're not dominos falling dumbly. Our decisions today may have been years in the making and bring together entirely disconnected influences—something your fourth-grade teacher once told you might intersect with an old song lyric and the pain in your elbow that you really need to call your doctor about. Two people, or the same person at two different times, can react to the same situation in opposite ways.

Once we do decide to act, intelligence gives us power out of all proportion to our raw bodily strength. By planting a lever in the right place, you could move a world (literally: in the fall of 2022, space scientists deflected an asteroid by hitting it in the right spot with a space probe).[34] That ability to apply power cleverly and selectively is also what differentiates us from inanimate physical systems.

Ismael argues that the twistiness of our deliberations makes us genuine features of reality, not reducible to our atoms, and justifies speaking of free will. She wrote in 2016: "I am a product of my past, but not a *mere* product of my past. . . . The universe is not just a flat landscape in which one thing happens and then another; there are special little causal hubs built to collect influence from across the landscape and filter it through a decision process that guides behavior. These little hubs are human minds."[35]

8

TIME AND SPACE

I HAD TROUBLE GETTING UP in the morning for school when I was a teenager. Once, I stumbled into the bathroom to get ready, only to find myself skydiving out of a plane. The landscape stretched out before me—what a view! I pulled the rip cord and gently glided toward the ground. It was exhilarating. Then I hit the floor of the bathroom and woke up.

Of all the ways our brains filter our perception of reality, time may be the most distorted of all. The second or two it took for me to collapse to the bathroom floor felt like entire minutes in my dream. The COVID lockdown furnished lots of new examples of such distortions. Days raced by for some people and dragged on for others.[1] In retrospect, the entire first year of the pandemic felt to me as if it had been erased from the calendar. I had visited Tucson just before it all started, and a year and a half later I still felt as if I'd just been to Tucson.

Once you become attuned to the brain's temporal trickery, you can find it everywhere. If you look at your right eye in the mirror, then at your left, you never see your eyes move—the brain hides that motion from you. If you look at the second hand of an analog clock, it may seem stuck at first, an illusion known as chronostasis, perhaps because of how the brain compensates for your eye movements. If you find yourself outraged by an offside call during a soccer match, which requires comparing activity in multiple places at the same instant, you realize how fraught the idea of "the same instant" is.[2] Our brains sometimes perceive simultaneous events as sequential, or sequential events as simultaneous.

Speech and music involve a kind of mental time traveling. If you transcribe an audio recording, you'll notice that speech is ambiguous or even unintelligible at the moment it is spoken—you don't know whether the word is "steak" or "stake" until you hear the qualifying phrase "on the grill."[3] Likewise, we perceive musical notes as part of a longer riff. The funny thing is that you don't feel any delay in comprehension; you feel you hear each word or note immediately. The brain hides its processing lag from us—up to a point. If a video soundtrack is synchronized within eighty milliseconds, it looks fine, but beyond that the video and audio become abruptly and maddeningly disjointed.[4] At big outdoor concerts, if you're near the stage, the music is in time with the jumbotron video, even though sound travels more slowly than light, but farther back the brain can't sync them up and the video looks like amateurish dubbing.

Our perception of space, too, is like a funhouse mirror. People judge distances asymmetrically: it feels farther from a landmark to your house than the reverse, and we exaggerate nearby distances and collapse far ones. The famous *New Yorker* cover that compressed the entire world beyond the West Side Highway to a narrow strip expressed a general human tendency, not a specifically New York parochialism. And these distortions carry over into more abstract forms of spatial awareness. We're "close" to our friends and so recognize them as distinct individuals, while thinking of strangers as all the same.[5] Liberals think conservatives move in lockstep and vice versa.

Because time and space are so central to consciousness, explaining how we experience them may also help to crack the broader puzzle of why we experience anything at all. Unlike most other experiences we have, our sensations of time and space can be analyzed.

The various illusions and distortions of perception help with that analysis—they are not failings of the system, but glimpses into its mechanisms. Once researchers articulate the more primitive feelings and judgments that time and space involve, they may be able to map them to brain activity, providing the elusive link from first-person experience to third-person understanding. "We have introspective access to spatial experience, which we don't have to the blueness of blue," the neuroscientist Giulio Tononi told me. "Blue is blue. You have nothing else to say. I can't dissect with my introspection the way blue feels. But I can dissect space."

MAKING TIME AND SPACE

If it's any consolation, physicists have as much trouble with time and space as our brains do. As they go looking for a fundamental time, it slips through their fingers. Time is directional. It flows; it passes. The present is real, the past fixed, the future open. Each moment brings something new. But none of this appears in physics. Time is simply a little *t* in our equations; it's just a label. Together with spatial position, it lets us place events into a causal sequence, but it doesn't differentiate past from future, it doesn't single out a present moment, and it doesn't flow. The universe does not unfurl, but simply *is*—it is laid out in its entirety, past, present, and future, like a landscape. It certainly doesn't look like that to us, though. We can't peer out and see our past, or backtrack to undo the mistakes of our youth.

As if the austerity of physical time were not enough, some of our most fundamental theories—those that try to unify quantum mechanics and the gravity theory of general relativity—don't even have a *t* in them anymore. Time at a fundamental level is not just too skeletal. It's not even there. In fact, the idea that reality is ultimately timeless dates to the ancient Greek philosopher Parmenides.[6] None other than Socrates said it went "over the heads of the rest of us,"[7] and physicists today feel the same way. Without time, how are they supposed to explain a world in flux?

Intriguingly, this problem of time sounds a lot like the problem of causation I brought up in chapter 7. The universe at root appears to have no concept of causation, and it seems it may not have a concept of time, either. Like causation, time may arise at a higher level, much as life and mind arise through the collective dynamics of matter.[8] But some of the familiar qualities of time aren't found in physics at all, and physicists scarcely talk about them; they assume the explanation must lie in brain science. "Physicists just basically would say it's an illusion, and then what they're doing is they're taking the problem off their desk and put[ting it] onto the desk of somebody else," said Craig Callender, a philosopher of physics at the University of California (UC), San Diego. "So it goes onto the psychologists' desk. But no one ever told the psychologists it was on their desk. . . . It's not being explained by anybody. It's crazy!"

Apart from being intellectually unsatisfying, this game of academic hot potato leaves big holes in physics itself. As an empirical science, physics is supposed to explain our observations, not deny them. Indeed, if it scorns the evidence of our senses, on what basis should we trust physics to begin with? Without a full accounting for time (even if time is an illusion), physics is self-undermining—another case of empirical incoherence.[9] To achieve that accounting, disciplines have to talk to one another.

Not everyone agrees that time is emergent. Lee Smolin, a physicist at the Perimeter Institute for Theoretical Physics in Canada, thinks our experience of time is closer to its fundamental nature than his colleagues generally suppose. He thinks there's more to it than a bloodless little *t* and that time really might flow after all. "The world exists as a sequence of things being made real or made definite," he told me. In support of this view, Smolin points out that *something* has to be fundamental. Usually physicists assume that their laws

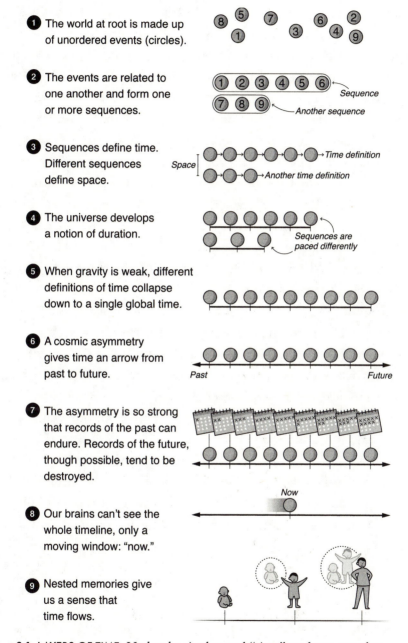

1. The world at root is made up of unordered events (circles).

2. The events are related to one another and form one or more sequences.

Sequence

Another sequence

3. Sequences define time. Different sequences define space.

Space

Time definition

Another time definition

4. The universe develops a notion of duration.

Sequences are paced differently

5. When gravity is weak, different definitions of time collapse down to a single global time.

6. A cosmic asymmetry gives time an arrow from past to future.

Past   Future

7. The asymmetry is so strong that records of the past can endure. Records of the future, though possible, tend to be destroyed.

8. Our brains can't see the whole timeline, only a moving window: "now."

Now

9. Nested memories give us a sense that time flows.

8.1. LAYERS OF TIME. Under the single word "time" we lump together a complex and only hazily understood set of ideas from physics with neuroscience. Gottfried Leibniz and, more recently, Carlo Rovelli have sought to describe how time's properties might emerge one by one.

are fundamental, and that time is in turn derived from them. But those laws seem arbitrary. If you asked, "Why those laws?" physicists would answer: "Because." If, instead, time is a rock-bottom feature of reality rather than a function of physical laws, Smolin suggests, then you have a hope of explaining those laws as the outcome of a process that unfolds in time.[10] In that case, we'll need to revisit much of what physics says about time.

Whenever a physicist talks about time, space can't be far behind. The two are conjoined twins—two aspects of the space-time continuum. Together they are the venue for everything that happens in the universe. Theory indicates that space falls apart at a fundamental level, as is most noticeable when scientists try to make sense of black holes. Space appears to be emergent—the product of the collective organization of the universe's building blocks.[11] Even Smolin agrees that space is emergent.[12] It's not often that physicists agree on anything, so their consensus that space derives from nonspatial physics is striking.

Although our experience of space isn't at odds with physics in the way that our experience of time is, physicists and neuroscientists have much to say to one another about it. Our mental experience of space, like the physical continuum, is emergent: our brains create a little inner universe from sensory information. Is this mental emergence process analogous to the one that may happen at the fundamental level of physics? "There is obviously an interesting parallel here," Tononi said. In one hint of a connection, AI researchers have found that theories of emergent space help them to make sense of neural networks; that is one of the many topics for us to explore in this chapter.

IN SEARCH OF LOST TIME

My writer colleague and friend Amanda Gefter spent months at an archive in Philadelphia reading the physicist John Wheeler's journals, coming as close as anyone has to living in the head of one of the twentieth century's most creative physicists. She discovered that on January 27, 1967, while staying at the New Yorker Hotel, Wheeler wrote that he and his coauthor Bryce DeWitt had found a "time bomb." In his wonderment, Wheeler alluded to Keats: "A fantastic story. We didn't dream it up. . . . Story *crying* to be told. . . . Silent, on a peak in Darien."[13]

The story goes back, like much else, to Einstein. His special theory of relativity famously showed that time does not tick at the same rate for us all. Each of us lives by our own clock. To me, yours runs slow, and what is now for you could be in my past. Less well known is that in his general theory of relativity, which provides our modern understanding of gravity, Einstein squeezed the last drops of reality out of time. He built the theory on the principle that there can be no external or God's-eye view of space and time—no absolute structure. That includes no fixed notion of time. To say an event occurs at a given time has no absolute meaning, and if the value of *t* changes, that has no meaning, either. The passage of time in his theory just reshuffles the deck chairs. It makes no difference to any observation we can make.[14]

The slipperiness of time was hard to see in Einstein's original equations, but became obvious when physicists in the '50s and '60s recast those equations to try to merge them with quantum theory.[15] If time in relativity is fluid, time in quantum theory is rigid. These theories are like two friends with opposing political views, who, to remain friends, agree never to discuss politics. That's why the *t* disappeared when Wheeler and DeWitt fit the theories together. The equation they derived for a quantum description of space was static, unchanging in time.

Wheeler's time bomb has a special status in physics and philosophy. It is known as not just a problem with time, but *the* problem of time. To this day, researchers disagree on how to solve it. But in some way time must arise from within the universe; it is not an externally imposed framework. A supposedly atemporal world surely seems temporal to those of us embedded in it—a case of the inside/outside problem.

In 1935 Einstein suggested with his usual prescience that "'time' is only a possible viewpoint, from which the other 'observables' can be considered."[16] DeWitt fleshed out this idea in a paper published not long after Wheeler's journal entry.[17] The universe as a whole may be timeless, but by splitting it into pieces and calling one piece the "system" and another the "observer," DeWitt showed that the observer will perceive time. The system changes in a relative sense.

One way to make sense of how time can emerge from timelessness was developed in the '80s by two other theoretical physicists, Don Page and William Wootters.[18] They imagined carving out a clock from the universe; even a single particle would do, for what is a clock but a little chunk of the universe that is correlated with the rest? By design, our mechanical gearboxes or oscillating electronic crystals match Earth's rotation (almost: it takes twenty-three hours fifty-six minutes for Earth to spin on its axis, but twenty-four hours for the sun to return to the same point in our sky, because Earth has moved in its orbit in the meantime). In Singapore, for example, when the clock reads twelve o'clock the sun is either directly overhead (noon) or directly underfoot (midnight), and at six o'clock the sun is either rising or setting.

Page and Wootters added a twist: Theirs is a Schrödinger's clock that reads twelve o'clock and six o'clock and every other time, all at once. The rest of the universe, too, is in a superposition—the sun is overhead and underfoot and rising and setting, all at once. But this clock, like all our clocks, is still correlated with Earth's spin. When the clock reads twelve o'clock, the sun is where it's supposed to be, and likewise for six o'clock and other times. Having set up this little multiverse of all coexisting possibilities, Page and Wootters asked what it looks like to an embedded observer. From the outside, the system never changes: it stays in its superposition. But from within, it evolves normally: an observer who looks at the clock sees some specific time. In 2013 a team of experimenters, led by Ekaterina Moreva at Italy's Istituto Nazionale di Ricerca Metrologica in Turin, demonstrated Page and Wootters's mechanism. They used a variation on Wigner's experiment from chapter 5—the one in which a superobserver observes an observer. The "observer" (represented by a photon) watched a clock (also a photon) and, from its perspective, time passed as usual. Meanwhile, the "superobserver" (a human) measured the combined observer-clock system. That person saw the observer plus clock in a vast but unchanging superposition, like frames of a movie all laid out. The setup confirmed that what is static from the outside looks dynamic on the inside.[19]

This scheme hinges on the correlation between the clock and the rest of the universe. Time thus fits nicely into a larger project of conceiving of the world as a web of relations, as Carlo Rovelli and other physicists have argued for.[20] "I think there's a clear way of getting time out of a timeless theory," Rovelli told me.

Normally we take time to be a steady cosmic drumbeat that drives all change. A pendulum swings, a clock ticks, a heart beats once per second. But if time doesn't exist at a fundamental level, all we have are relative changes. A pendulum swings once per clock tick or heartbeat. Practically, we can't possibly map out all these relations, so we use time as a shortcut, careful to remember that it is our construct. In the first of his short, delightful, elegant books on science, *What Is Time? What Is Space?*, published in English in 2006, Rovelli wrote: "Time is an effect of our ignorance of the details of the world. If we had complete knowledge of all the details of the world, we would not have the sensation of the flow of time."[21] On this view, time plays the same role in our lives that money does in an economy: it provides a convenient medium of exchange and has no value on its own. In principle, we could shred all our paper bills and get by on gift exchanges or barter transactions.

A common currency presumes commonality; the challenge is

to explain it. In other words, we need to understand why beating hearts, swinging pendulums, and spinning globes keep pace with one another, so that we can introduce time to describe them. We take the mutual consistency of natural processes for granted: Apart from the slight deviations predicted by relativity theory, once we set a clock, it stays in sync with other clocks and with the rotating Earth. We can rely on clocks to help us bake a cake or catch a train. But their synchrony isn't automatic. In principle, the universe could have been arrhythmic, with hearts, pendulums, and planets all dancing to their own beats, failing to sync up. Rovelli has explained this synchrony in terms of the same collective behavior that gives rise to heat—things in the universe interacting and organizing themselves to create the necessary correlations.[22]

There is an intriguing analogy between how the universe constructs time and how our brains do: it may be that we don't need to have an innate notion of time, but can extract it from the change we perceive in the world. A team of neuroscientists—including Warrick Roseboom and Anil Seth at the University of Sussex and Zafeirios Fountas, who recently left academia to develop AI at Huawei Technologies—has sought to understand time perception using predictive coding, one of the physics-inspired theories of consciousness that I described in chapter 3.[23] If the world surprises us—something happens that our brain didn't predict—we tend to notice and remember that. The team's idea is that these surprises are the ticks of our subjective clock. "There isn't a fundamental grid underlying it," Roseboom said. "It's purely constructed out of salience." An eventful scene seems to last longer. The team demonstrated this using an artificial neural network that was shown videos of scenes ranging from quiet cafés to busy streets. The network was prone to many of the same temporal illusions as we are. Asked to estimate how long ten seconds on a busy street lasted, people will say thirteen seconds, whereas the same interval in a quiet café feels like seven seconds; the network made similar misjudgments. The misjudgments depended on whether people were in the thick of things or looking back. A vacation flies by at the time, but seems so long when you're telling people about it afterward.

Ideas such as Rovelli's relational time are deep and still being developed. The important point is that, even at this most basic level of reality, time may be connected with our own existence. To be sure, fundamental time doesn't require human minds. The universe fractured into subsystems long before conscious creatures ever walked the Earth, and the meshing of those subsystems implied a notion of time. Nonetheless, time, life, and mind share certain preconditions: All require that an initially formless universe become organized and differentiated. And like conscious experience, time exists only from an insider's perspective.

The universe is like a book: motionless—only good for pressing flowers or raising computer monitors—until you enter into it and are swept along by the narrative.

TIME'S ARROW

Why don't we remember the future? Most people would answer, Because it hasn't happened yet. But for physicists, the future is out there right now as surely as stars and galaxies are—and yet for some reason we see it only in due course.

The directedness of time—the arrow pointing from past to future—is the one aspect of our temporal experience that physicists think they have a handle on. By general consensus, it is a property not of time per se, but of matter within time. "Future" and "past" are like "up" and "down"—labels that mean nothing to space itself. To us, "up" means away from Earth's center, and in absolute terms what is up to someone in Singapore is down to someone in Quito.[24] "Future" has an analogous significance. "In fundamental physics there is no unique time variable and there is no preferred direction of time," Rovelli told me. "These emerge in the specific contexts in which we are."

The laws governing individual things are arrowless. If the cue ball rolls down the table, strikes the eight ball, and sends it rolling, that seems like a directed process. But you might as well say the eight rolls up the table and strikes the cue. Based solely on the movement of the balls, it is ambiguous whether the collision is in a ball's future or past. Only when you have a group of things do you have the potential for an orientation. In the opening break shot, the cue ball scatters fifteen object balls. The reverse process, in which fifteen balls converge on a triangular area, all stop, and eject the cue ball toward the head of the table, is wildly unlikely, because it requires such a high degree of fine-tuning. It's not strictly ruled out, but please email me if you've ever seen it occur in regular play. Such exactitude typically requires highly controlled physics experiments to achieve.[25]

Physicists describe this collective behavior using the concept of entropy, which they devised in the nineteenth century to study heat, but now apply to all systems: balls, molecules, polymer chains, gravitating masses, computer bits. In their books and articles, physicists usually refer to entropy as a measure of disorder, without clarifying what "disorder" means (teenagers and parents can look at the same bedroom and disagree on whether it qualifies as disordered). A more direct translation of the mathematical definition of entropy is hidden complexity—the number of ways a system could be internally structured while presenting the same face to the world. The connection between two levels of description is essential to the concept of entropy. If you are keeping up a running commentary on a game of pool, you might use summary or higher-level terms: the balls are "racked," the balls are "scattered." There are fairly few ways for balls to be properly racked—they have to fit inside a triangle with very little wiggle room. But there are countless ways for them to be scattered. So the word "scattered" hides a lot more complexity than the word "racked," and we say that a system of scattered balls has higher entropy than a system of racked balls.

For no other reason than the relative numbers of possible configurations, it's much easier for the balls to go from racked to scattered than the other way around. If the balls are already scattered, they will probably stay that way, short of someone picking them up by hand; they're very unlikely to return spontaneously to the racked, low-entropy configuration. As long as the balls were originally racked—the starting point is crucial—this natural tendency gives the collective a sense of progress that the individual balls lack. This principle applies not just to pool games, but to the observable universe as a whole, whose starting point is the big bang. The entropy of the universe can increase only if it was originally much lower. The arrow of time ultimately reflects an asymmetry in the initial conditions of the universe.

Every directed process—from stirring cream into coffee to your life path from birth to death—reflects this cosmic asymmetry in miniature. Our daily life is directly governed by the nature of the universe on its largest scales. "Every time you break an egg, you are doing observational cosmology," the cosmologist Sean Carroll has written.[26]

But while the entropy explanation for the arrow of time is compelling, it is incomplete. Cosmologists still aren't sure why the universe started off with such low entropy. One intriguing possibility is that it has to do with how we define the levels of description. If you decide the higher-level states are racked versus scattered pool balls, whole versus broken eggs, and diffuse versus clumped cosmic matter, then you see a clear progression from one state to another and therefore from past to future. But who says these are the relevant categories? These are not absolutes; they depend on how the system looks to us.[27] "This time arrow depends on our coupling with the rest of the world, hence is perspectival," Rovelli told me. For him, there doesn't need to be any explanation for the cosmological initial conditions, because no matter how the universe was set up, there would always be some way to define higher-level states according to which entropy increases and time has an arrow. All directional processes, including biology, cognition, and consciousness, make use of these states. By Rovelli's logic, all intelligent creatures in the universe will see an arrow of time. They needn't see the same arrow that we do, though, because they may be interacting with different families of high-level states. For some, the arrow might point the opposite way as ours, so that our past is their future.

Most physicists and philosophers think the physical arrow of time entails the psychological one—that entropy growth accounts for why we don't remember the future. But it's not a given that the psychological arrow would follow the physical one. Although the biological processes that are involved in memory formation depend on the arrow of time, physicists need to be more specific about which properties of memory would tether it to an entropic gradient.

In 2004 Len Mlodinow of Caltech and Todd Brun of the University of Southern California argued that the salient property of memory is representational flexibility, or what they call generality.[28] If I have tea this morning, I form a memory of having tea. If I have coffee instead, I form a memory of having coffee. If I can't make up my mind, I form a memory of indecision. The memory system can handle all these options. It thus needs to be sensitive, but not too sensitive. It should adjust to changing circumstances without failing entirely.

This is where entropy comes in. To see why, go back to the pool table. In the opening break, the cue ball scatters the object balls, and variations on the shot will lead to different but basically equivalent outcomes. If you view this process in reverse, the object balls all have to approach a single spot just so: any deviation will lead to an utterly different outcome—namely, the balls won't come to rest in their triangular starting configuration, but will scatter. The forward break, in which entropy increases, is like a memory that can store multiple options, whereas the reverse, with decreasing entropy, is like a memory that can store only one option. This implies that a memory worthy of the name—one that can handle many options— requires entropy to be increasing overall. To be sure, Mlodinow and Brun's argument requires that entropy be increasing not just overall, but locally, inside the brain. It is not guaranteed that the brain is aligned with the general arrow of the universe.[29]

One fascinating consequence of this theory of memory, if true, is that a record of the future would in principle be possible. The arrow of time is a statistical effect, so what seems like impossibility is really just improbability. Nothing stops pool balls from spontaneously returning to their starting triangle, and nothing prevents the world from bearing the traces of events that are yet to happen. They would manifest themselves as bizarre reversals of the usual entropic tendencies, as in the mind-bending science-fiction movie *Tenet*, which has fun with examples such as objects "falling" up. Ordinarily, we release something and it falls to the floor, dissipating its energy as heat. In *Tenet*, heat converges on an object and ejects it upward into your waiting hand. Bullets fly out of targets back into gun barrels and are stopped by the firing mechanism. Presumably we don't see such reverse-entropy events happen in real life because of the incredible fine-tuning that would be required to set them in motion in a way that we'd recognize. In the case of the upward-falling object, the heat pulses have to converge exactly right; otherwise, instead of seeing an object fall up, we'll just notice some random heating.

Some researchers go further and say that, by our very nature, we could never see signals from the future. Huw Price, an emeritus professor of philosophy at the University of Cambridge, has argued that our senses and instruments just aren't set up to detect them.[30] Normally, light enters our eye and nerve signals flow to our brain. Reverse-entropy light would be a wrong-way nerve signal that causes the eye to emit light. Even if this were physiologically possible, our brain wouldn't perceive this process as an intake of information.

In addition, the physicist Lorenzo Maccone of the University of Pavia in Italy has argued that we couldn't form a memory of a reverse-entropy process.[31] Our memories are not passive impressions of events in the past. They form when we interact with the world, and in that interaction we become quantum-entangled with whatever we remember, as I mentioned in chapters 4 and 5. My father really does live on in my memory—this is not merely poetic license, but a genuine physical linkage between neural activity in my brain and events long past. But Maccone notes that memory formation requires that neural activity and events be initially unentangled. So he suggests that if entropy were reversed and those events were undone, our memory of them would be wiped, too.

There's one tantalizing exception to the impracticability of remembering the future. Price and others have speculated that the arrow of time gets bent in experiments involving entangled quantum particles. As I brought up in chapter 5, you can create two such particles and separate them, and they will remain synchronized, responding in matching ways to a wide variety of experimental conditions. Their synchrony seems mysterious, but these researchers argue it would be entirely natural if the particles had foreknowledge of what those conditions would be.[32] For instance, suppose you create a pair of entangled photons at noon, send them off in different directions, and, at one o'clock, place polarizing filters in their paths, choosing randomly among several available filters. If not for quantum entanglement, you'd expect the photons to behave independently; if one passes through its filter, the other has a fifty-fifty chance of passing through it. But with entangled particles, if one passes through, the other always also will.

Price and others suggest the reason is that the information about your choice of filter at one o'clock propagates back in time to noon, affecting the creation process so that the photons are tailor-made for the filters they will encounter. That way, if one passes through, the other also will. In short, the experiment creates a situation of retrocausality, in which the future can discernibly influence the past. "My view is that retrocausality, or the influence of future boundary conditions, already shows up in well-known quantum-mechanics experiments," Price said. To be sure, this is just one of many interpretations of those experiments; and even if it were true, the premonitions would be extremely limited—far too restricted to give you ESP, for example. The universe is set up in such a way as to cloak from us what is to come. Perhaps this is for the better.

The essential point is that the arrow of time, rather than being a fully objective feature of the physical universe, depends, in part, on our own constitution as conscious agents. The past and future have no fundamental meaning in nature; they are defined only by the asymmetries of our own knowledge and agency. The past is that which we cannot change; the future, that which we know not of.

DON'T LIVE IN THE PRESENT

The arrow differentiates past from future, but what about the present? The present has no place in physics at all. How could it? Every moment feels like the present when you're in it. "Now" is an indexical concept, like "here" in chapter 6—not part of a third-person description of nature, but extra information that is specific to you. Furthermore, in physics, the present moment is at most a value of the t variable, and your experience is much richer than that: not just an infinitesimal slice of history, but a thicker slab. Moments are not instants.

Experiments show that the shortest interval we can discern is about thirty milliseconds. Clicks or flashes separated by less time than this appear to be simultaneous even when they occur one after the other. We hold in our heads a somewhat longer period, a moving window of a few hundred milliseconds, perhaps stretching to several seconds.[33] If something moves a discernible distance within a few hundred milliseconds, we can tell; otherwise, it looks fixed to us. Rick Grush, one of Craig Callender's philosophy colleagues at UC San Diego, offered an example from John Locke: On an analog clock, you can watch the second hand sweep around, while the hour hand looks stuck. You can tell the hour hand is moving by looking back at it every so often, but that's a very different judgment.[34] "I can deduce that it's moving, but I can't see it move, whereas with the second hand, you can just see that thing move," Grush said.

Try this elaboration on Locke's observation. Set up an analog clock with a second hand (one that glides continuously, rather than ticking in little jumps), and then step back from it. See how far away you can get and still perceive the second hand moving. I find that, from across a room, the second hand looks uncannily as though it's both moving and standing still, like a series of fast snapshots. Viewed from that distance, the second hand moves by such a slight angle within the window of perception that my eyes don't have the resolving power to see the motion. I've also noticed this in classrooms—if you sit at the back, the second hand of the clock on the front wall doesn't seem to move. I wonder how this affects students' perception of class duration! Humans have come up with all sorts of tricks, such as time-lapse photography, to compensate for the narrowness of our direct awareness of motion.

Psychologists refer to the felt duration of "now" as the "specious present." It's "specious" because the present is not present; what we experience as "now" is already largely past. In one of the many minor mysteries of this subject, William James in 1890 credited the coining of the term to one "E. R. Clay," and for over a century no one knew who that was.[35] Clay had published his book coining the term anonymously, so James clearly had some inside information.[36] Only fairly recently did Grush and fellow philosopher Holly Andersen do some detective work and identify him as E. Robert Kelly, a cigar maker who took up psychology in his retirement.[37] A note corroborating his authorship had passed down through the family, proving that you should never throw away your grandparents' keepsakes.

The specious present is where subconsciousness shades into consciousness. Our subconscious handles basic perception and quick, direct responses. Anything longer engages the conscious mind.[38] The subconscious, for example, recognizes individual words, but can't string them together into full sentences. Even when you're lying on the gurney under heavy sedation, neurons are firing in response to what the surgeons and nurses are saying, while the brain areas associated with comprehension are unresponsive, so you are unlikely to remember the conversation.[39] The temporal extendedness of consciousness may be why animals evolved it. If engineers need a practical reason to design consciousness into their AI systems, the advantages of temporally extended processing could be it.[40] For instance, they have been training language-processing systems such as ChatGPT to store the intermediate steps in their reasoning, giving them a form of working memory that helps when solving math and logic problems.[41] David Chalmers has suggested that such a capability could one day justify considering these systems sentient.[42]

In the mid-2000s Grush sought to explain how our brains construct the specious present; he proposed a version of predictive coding.[43] His model was inspired by experiments showing that the specious present extends not just into the past, but into the future. When you ask people tracking a moving object to point at it, they consistently point slightly farther ahead than it really is. "The perceptual system is not just telling people where things are, but is actually giving them at least a little bit of prediction into the future," Grush told me. Because your brain is actively forecasting, you experience what has not yet happened. You can catch a ball because you literally see it where it will be.

For Grush, the specious present not only lets the brain get a jump on events, but also gives it time to sync up nerve signals arriving on different schedules from all parts of the body. "Eighty to a hundred milliseconds is about the length of time that, for humans, is the longest delay in the proprioceptive input," he said, referring to signaling within the body. "So how long does it take for a signal to get from my foot, the farthest thing from my brain, back to my brain?" If the brain sliced time finer than one hundred milliseconds, we'd experience simultaneous events as occurring in succession. We might touch our toe with our finger and feel it first in our finger and then in our toe, rather than in both at once.

Advocates of integrated information theory (IIT) have their own explanation of the specious present. They think the duration of the present is delimited by how long it takes the brain to process sensory inputs. The visual cortex, for example, is a multilayered neural network whose different layers hold snapshots taken at intervals over a period of hundreds of milliseconds. "They're not representing the same instant in physical time," said Andrew Haun, an experimental psychologist at the University of Wisconsin–Madison who specializes in applying IIT to particular categories of experience. "But you're experiencing them all together, and that's the window over which temporal experience exists."

Physicists tell us that the universe is laid out in time as surely as it is laid out in space. We do not see all of time spread out in front of us, but we glimpse a timeline lasting a few tenths of a second. "That specious present is the equivalent of the canvas of space," Tononi said. Drawn on this temporal canvas are the syllables in a word or the notes in a melody. Each moment may be fleeting, but is still long enough for us to inhabit. Everything we feel, we feel in this short interval.

GO WITH THE FLOW

Some aspects of time we observe directly: a sticky clock hand makes us aware of the specious present, and breaking an egg—and realizing we can never unbreak it—betrays the arrow of time. But other aspects of time are hazier. People often say that time flows like a river, sweeping us along. Events approach, time slips away, the past recedes: time has a kind of motion to it. Yet we don't directly see any such movement.[44] "Things don't look flowy to me," Callender said. "People say they have this flow sensation, but they're terrible at saying what the heck it is." And if you stop to think about it, the idea of the flow or passage of time can seem downright silly. If time is the measure of change, it can't itself be changing. Flow implies velocity—a change of position over time. How fast could time be flowing? One second per second?

Callender suggests that the passage of time is less a visceral feeling than an abstract judgment. It incorporates some elements of physics—that we remember the past, not the future, and that events can play out at different rates for different observers—and some elements of cognitive psychology, such as how we form and conjure up memories. And it is also a product of self-reflection. We start talking about time passing when we become wistful about change: when we wish we were still on vacation or are weirded out by dropping off a child at college.

In 2004 the physicist James Hartle of the University of California, Santa Barbara, ventured an explanation for the flow of time. To do so, he repurposed a simple model of cognition that he and the Nobel-winning physicist Murray Gell-Mann had developed in the late '80s to study a separate problem: quantum measurement.[45] It was an upgrade of Hugh Everett's model of the observer as an eye plus memory. They called their model an IGUS, for information gathering and utilizing system. Its key feature is that, unlike the systems that physicists typically study, it is not merely reactive. An IGUS takes its past experiences into account when plotting a course of action. It continually updates its memory as it takes in new information. And that, Hartle suggested, gives the IGUS a sense that time flows. For the IGUS, what is "now" is always changing because its memory is always changing.[46]

Other theorists have also taken up the origins of time's flow. The philosopher Jakob Hohwy of Monash University in Australia and his colleagues have argued, based on predictive-coding theory, that it is not the updating that makes time appear to flow, but our awareness of this updating. The brain predicts that its prediction is provisional and therefore likely to change.[47]

Callender, for his part, thinks just laying down memories is not enough to give us the sense of flow.[48] Memories, on their own, he points out, are disconnected snapshots; they mark a change of experience, not necessarily an experience of change. For us to sense that time flows, our memories need to be linked together—we need to remember remembering. Not only do I recall what happened yesterday, but also I recollect that yesterday I was showing my friends pictures of the jazz festival I had gone to, and that the previous day I was reading a review of a movie I'd seen the night before. These linkages among multiple events in my past are spotty, but there are enough of them to give me a sense of continuity. "At each point I say, 'This Craig comes from that other one,'" Callender said. "Why? I've got nested memories: I've got memories of memories."

Personal identity, Callender suggested, is responsible for our impression that time flows. From our nested memories, we know that we know more today than we did yesterday, and more yesterday than the day before, so it's evident we're changing. But we don't feel our essential selves change. We view the self from the inside; this is our natural reference frame. Wherever you are, you are here, and whoever you are, you are you. If we could gain some distance from ourselves, we would clearly see we are not, in fact, the same people we were a moment ago, let alone when we stole that street sign in high school. But once we commit to holding our identity constant, we displace our personal transformations onto time, Callender argued. We feel we are stationary and time rushes past, like sitting on a train and sensing that the countryside is sweeping past you rather than the other way around.

There's an interesting tension in the psychology of time's flow. To feel that time passes is to feel that the world is changing. But this feeling is produced by continuity rather than by change: We can become aware of change only because we persist through it. People can get melancholic about time's passage, since it makes us feel that life is short, and yet time's passage also reveals to us our own staying power.

Once we get to these highest-level puzzles of what time means, they fall more on neuroscientists than on physicists. Time is as much a reflection of the workings of our minds as of the fundamental nature of the universe. "Kant thought that time was all in the mind," Callender told me. "For me, I think modern cognitive neuroscience shows that Kant was right about a whole heck of a lot. . . . I would have started off way against Kant on that kind of thing. Learning all the cognitive science of time perception was really eye-opening—to see how much of us was mattering."

SPACELESS TO SPACE

In comparison to time, with its multiple layers and confounding mash-up of physics and psychology, space might seem easy. By definition, it's a whole lot of nothing. We see it right in front of us, so describing it requires none of the analogizing that goes into our experience of time. We don't have to say that it's kind of like a river or kind of like a dimension. But in physics, if something seems easy, brace yourself. One of the most remarkable lessons of modern physics is space's mind-boggling complexity. The seeming void acts like a material object that bends, warps, vibrates, heats up, and cools down. By broad consensus it is indeed a material object of sorts, made like any other out of some sort of "atoms"—fundamental building blocks, as yet unidentified. "If this perspective turns out to be generally true, it would be the most dramatic change in our view of spacetime since—god, I don't know when, the beginning?" said the theoretical physicist Vijay Balasubramanian at the University of Pennsylvania.

A weird thing about these "atoms" of space is that, if they are to generate space itself, they can't themselves be spatial—that would be circular. So they can have no spatial attributes: no size, no shape, no position; they're not little scraps of space.[49] To the contrary, they span all of space. What makes it so hard to develop theories of the origin of space is that physicists have to go cold turkey on the geometric notions that have defined their field since Democritus.

A major research program now seeks to explain spatial attributes as correlations that the "atoms" develop with one another.[50] The physics is highly mathematical and not easy to interpret, but the emerging picture is something like the following. The Andromeda galaxy is 2.5 million light-years away, but the space it sits in is made of the same "atoms" as our solar system, merely in novel combinations. At a deep level, Andromeda and the solar system may be sitting right on top of each other, but appear far apart because they are only weakly correlated. In effect, the universe is not built from the bottom up, out of localized parts that glom together, but from the top down, from a whole that fractures like shattered glass. The whole precedes the parts.

Physicists' attention to space is spilling over into neuroscience. Because our experience of space seems pretty blah—it has no color, no texture, no emotional associations—most neuroscientists and philosophers have until recently given it little thought. "They usually refer to the redness of red, the painfulness of pain, the beauty of the sunset—you name it," Tononi said. "Anger, emotion, love, smell:

they all talk about those, which is fine; they certainly are part of consciousness. They miss the number-one, the biggest, component of every experience we have, which is it feels extended."

Once you begin to think about it, though, space is every bit as rich as any other experience we have. "The structure of experienced space is crazy," Tononi said. "An empty canvas or the dark sky is immensely structured." Consider the experience of playing chess. It's all too easy to scan the board, make a move, and immediately regret it because you missed a threat that was right in front of you. Ranks, files, diagonals, Ls: a mere sixty-four squares have too many spatial relations for most people to keep track of. Even grandmasters are known to miss what's right in front of them.[51]

Not only are neuroscientists learning from physicists not to take space for granted, they are also encountering theoretical problems concerning our perception of space, such as circularity—having to construct experiential space without presupposing it—that tantalizingly parallel problems in physics. Our brains create little maps of space in our heads,[52] but those alone can't account for spatial experience. Just as red is more than a wavelength striking the eye, spatial experience is more than a layout of what's where. The neuroscientist Joanna Szczotka, who is now working toward her PhD at the University of Wisconsin–Madison, told me IIT is unique in trying to describe how the brain actively constructs a sensation of spatial extendedness. "Other theories simply delegate this problem to the external world—qualities are mostly 'out there' and we are simply representing them somehow, so there's not much to explain here," she said. But representation alone is not experience.

In another parallel to physics, some neuroscientists have argued that spatial experience emerges from the top down. Chess again illustrates the point. Through hard study and long experience, chess masters come to see pieces on the board not in isolation, but as full patterns. They take in swaths of the board, or even the whole thing, at a glance.[53] William James argued that our routine spatial experience

is like that. We perceive space first in its entirety—as elemental sensations of corporeality and situatedness—and then we break it down
into a granular structure. "The primary retinal sensation is a simple
vastness, a teeming muchness," James wrote. "The perception of
positions within it results from subdividing it. The measurement of
distances and directions comes later still."[54] The whole, again, precedes the parts.

SPATIAL QUALIA

Spatial experience is to consciousness research what the hydrogen atom is to physics: hard enough to be interesting, but easy
enough to figure out. There's a clear path to solving this problem.
We can look into ourselves, contemplate our own spatial experience,
and then judge theories of consciousness on whether they reproduce
its properties.

Andrew Haun told me of the personal factors that set him on his
path to studying spatial experience. While working as a postdoc in
Boston, he spent much of his free time doing Tae Kwon Do. Perhaps
because of the intensity of his practice, he had several distressing
neurological episodes. One was amaurosis fugax, a temporary loss of
the blood supply to his left retina. For several minutes, he saw mostly
blankness. "It was pretty creepy," he recalled. "If I closed my right
eye, I could see through a little window, because the fovea, the very
central part, has a different blood supply. But surrounding it, everywhere else, was this perfect, flat, gray emptiness. I'll never forget."

He also got visual migraines every week or so. I've had a few of
these myself. A little area in your visual field starts to shimmer like
a desert mirage. Slowly it expands in a wave, looking like the force
fields in *Star Trek* or *Dune*. At first the spectacle is actually kind of
cool, but it gets scary because behind the wave is a blind spot. You
find yourself peculiarly unable to see words on a page. You know

the words have to be there, but somehow they're just not visible. All you can do is to stop and lie down until it passes. "When I would have these auras, I didn't see a blank emptiness in the region of the blind spot," Haun said. "There was just nothing there. And that was before I came [to Wisconsin]; that's probably what started to put me onto these kinds of questions—how weird it is. It was like the space behind your head. You know it's there; you don't have any experience of it."

When this visual disturbance first happened to me, I went to see an ophthalmologist, who examined my retina and optic nerve and told me they were fine. The migraine had been produced by dysfunction in my brain's visual cortex, she said. With amaurosis fugax, you still perceive space even if it looks empty, but with a visual migraine, space is erased altogether. The brain, for a time, no longer constructs spatial experience. Haun's and my visual migraine episodes were fortunately temporary, but many stroke patients are permanently blinded to one side of their visual field. They do not see a flat, gray field there; rather, they do not perceive that region at all. An object there never enters their visual awareness, and if they handle or bump into it, they feel as if it had come from outside space.[55]

Tononi hired Haun in 2014, around the time Tononi was embroiled in a debate with the computer scientist Scott Aaronson.[56] Aaronson had complained that IIT gives false positives for consciousness. A computer memory—a simple grid of units interconnected to provide redundancy against data loss—could have a very high degree of information integration. The theory would say it's conscious. Surely it isn't, he asserted. Tononi retorted: What makes you so sure? Huge swaths of the brain, notably the visual cortex, are grid-like. "No matter what Aaronson or anybody else says—that of course grids are not conscious—you are mostly a grid, so you should pay respect," Tononi told me. In fact, when he and Haun worked through what kind of conscious experience a grid-like area of the brain might have, it roughly matched the human experience of space.[57] Our experience of space may be a direct consequence of the structure our brains evolved to process visual information.

Consider what space feels like to us. At its most granular level, it is a mosaic of localized places, which Tononi and Haun call "spots." You consider sets of spots and, from them, define geometric relations. The location of a spot is defined by what spots enclose it; its size, by how many spots it encloses; its distance from another spot, by the size of a set that encloses both spots. In short, Tononi and Haun turned space into a big, complicated Venn diagram.

The idea is to look for a structure matching this Venn diagram in the brain and other neural networks. Tononi offered the example of an extremely simple grid-like network modeled on the nervous system of roundworms: seven neurons in a row. Each is wired, with diminishing strength, to itself, to its neighbors, and to its next-nearest neighbors. This network has a hierarchy of relations. At the lowest level, you take each element in isolation. One step up, you take them two by two. Adjacent elements couple to each other strongly, so a pair is more the sum of its elements; in fact, such a pair is an entity its own right and plays a causal role above and beyond that of the elements that make it up. Distant elements, on the other hand, don't interact at all, so nothing is gained by pairing them.

In this vein, you continue to trios, quartets, and so on up to all seven neurons. Even this little network has quite an elaborate structure. Haun and Tononi showed that its internal relations match the sundry properties of space. Individual neurons are the smallest visual spots, whereas groups are larger, more encompassing spots; and their Venn-diagram relations define how the spots are ordered and how far apart they are. The system of all seven neurons, taken as a unit, is the entirety of a miniature space; it has an internal coherence—it's not just a string of atomized locations, but has a sense of extension. "The highest-order concept is the one that says, in words, it's all part of one big space," Tononi said. "This is all one thing, tied together."

The grids in our own brain are clearly much more sophisticated, but follow the same principle. In retrospect, it might seem obvious that grids could generate spatial experience, since they do look like space: a repetitive array of places where things could potentially reside. But not every grid captures spatial relations. Like the temperature of porridge, the connectivity has to be just right. With too few interactions, the grid doesn't cohere. With too many, the spots blur together. "If the graph is of a certain kind, then you can work out that it will be space-like in terms of its cause-effect power," Tononi said. "If it were connected differently—like all to all—it wouldn't be like space." This is the same happy medium required to give a high degree of information integration. What was a bug to Aaronson is a feature to Tononi.

Tononi and Haun are now extending their model to account for other aspects of experience. Suppose each unit in the grid were not a single neuron, but a cluster of thousands of them. If this cluster didn't have the right connectivity for space perception, we wouldn't perceive it as a part of space, but as some feature of each point in our visual field that we could not analyze introspectively—maybe color.[58]

SPACE ON THE BRAIN

A particularly fascinating analogy between emergent space in physics and spatial experience in the brain comes from theorizing about neural networks. Many physicists and AI researchers have argued that artificial neural networks (and perhaps natural ones) work on principles eerily similar to those of emergent space.

Consider what it means to move through the layers of an artificial neural network. In most image-processing systems, for example, the first layer handles pixels and edges, subsequent layers build up geometric figures, and upper layers detect eyes and fur. Depth thus corresponds to more complex and typically ever-larger features. Moving through the network is like zooming out. You start at the maximum zoom, highlighting every detail, and then pull back to take in the bigger picture.

Networks thus embody renormalization, the technique that physicists use to relate the multiple scales of nature that I mentioned in chapter 7.[59] You start with the laws of particles, consider higher-level structures, and work out what the laws of those structures should be. You can also start at a higher level and infer what must be happening on a microscopic scale.

Renormalization is not just a zoom dial, but also the underpinning of a leading theory of emergent space known as the AdS/CFT duality. The idea is that the zoom level is like a dimension of space. At the simplest level, if you point your camera at a scene and zoom in, it looks as though you have moved closer to the object; although the camera image is flat, the zoom level adds a sense of depth. In the AdS/CFT duality, the role of the image is played by a quantum fluidlike system that resides in a certain number of spatial dimensions (abbreviated, for technical reasons, as CFT), the renormalization procedure adds another dimension, and the full scene with depth is a universe with a special kind of geometry (abbreviated AdS). If you start with two dimensions, you end up with three. To shrink or enlarge a structure in the original 2D space is to move through this new third dimension. As with the camera image, all the depth information exists within the quantum fluidlike system and can be drawn out by viewing it the right way. It's a fascinating concept and my earlier books and those by the string theorist Brian Greene provide more details if you're interested.[60]

Putting two and two together, if networks are like renormalization, and renormalization is like emergent space, then networks must be like emergent space.[61] You can get a sense of the basic idea just by looking at diagrams of neural networks, such as figure 2.1. These typically show the layers of processing lined up like books on a shelf, with an arrow showing the flow of information from left to right. That arrow is a kind of spatial dimension. If you introduce a signal into a network, it propagates from one layer to the next as if moving through space. "This horizontal direction is the emergent direction of the geometry," said Koji Hashimoto of Kyoto University, a theoretical particle physicist who has explored the analogy between networks and space.

What justifies calling the horizontal direction "emergent" is that the intermediate layers in a network are not set up at the outset, but determined by training. Consider again a network that classifies pictures of pets. The input is a photo, the output is a label, and in between are multiple layers of neurons. You give it images and labels, such as "kitten" or "puppy," and the training procedure fills in the rest, strengthening or weakening the connections among neurons in order to connect input to output. "Your network is automatically tuned to map the left-hand side to the right-hand side, and so it's emergent inside," Hashimoto said. So it is with space in the AdS/CFT duality. You start with the quantum fluidlike system—it is like the input; and spatial relations derive from it—they are like the neuronal connections.

In 2018 Hashimoto and his colleagues demonstrated that the analogy is surprisingly tight.[62] They trained a neural network on simulated data from a quantum fluidlike system that they knew should give rise to a black hole. Apart from referring to a black hole, they did not give the network any further indication that gravity operated, let alone what its laws were. Yet the machine proceeded to predict a space that not only contained a black hole but also had a geometry that agreed with Einstein's theory of gravitation. So neural networks really do seem to capture some of the basic physics of space, an insight that theorists are still trying to make sense of. Just on the practical level, it gives them a much-needed new mathematical tool to study spatial emergence. Hashimoto is now cataloging the shapes of space that might emerge from diverse network architectures. "My intention is to find a geometry which was unknown to humankind," he said.

Some speculate that the resemblance of neural networks to the structure of space is no mere analogy. In chapters 1 and 2, I talked about how John Hopfield and other neural-network pioneers thought of networks as giant arrays of particles, not unlike crystals. It goes the other way, too: You can think of the particles that fill the universe as literally a giant neural network. All the dynamics of physics—particles moving around, exerting forces, and so on—can be thought of as the dynamics of a neural network that is being trained.

"Maybe you can reverse the problem—not just only use physics to understand how machine learning works, but actually use machine learning as a model of physics," said Vitaly Vanchurin, who was a physics professor at the University of Minnesota Duluth when I first talked to him, but who has since left to found the AI startup Artificial Neural Computing. Under some conditions, he said, a network might act according to quantum mechanics; under others, according to the general theory of relativity; such behavior would reveal the essential unity of those two theories.[63] A neural network also organically includes a notion of observers, thus offering a solution to the inside/outside problem. "There are three things that need to be unified," Vanchurin said. "We have to unify not only quantum mechanics with general relativity, we have to unify with observers. We have to understand how they emerged as well." Provocative though his idea is, it hasn't gotten much traction among most other physicists. "It's really an interpretation of existing physics rather than the discovery of some new connection," said Kyle Cranmer, who specializes in applying ideas and tools from AI to physics.

But the physicist Lee Smolin thinks that conceiving of the universe as a neural network might achieve his goal of explaining why the laws of physics are what they are. He, fellow theoretical physicist Stephon Alexander, the computer scientist Jaron Lanier, and their colleagues play off the close mathematical resemblance of neural networks and quantum theories of particles.[64] We frame them both in terms of matrices, which are giant spreadsheets of numbers. For a neural network, those numbers represent the neuronal interconnections and are set by training; for the particle theories, the numbers represent particle behavior and are set by—well, most physicists think they aren't set by anything; they just are. But Smolin and his coauthors think the numbers representing particle behavior may be set in exactly the same way as a network's numbers. "We map a machine-learning recognition problem to a universe choosing its laws," he said.

This isn't like learning to recognize cats, though. It's more like evolution in nature, which some biologists in recent years have conceptualized as a learning process.[65] Smolin and his coauthors suppose the laws of physics, like species, evolve over time and eventually reach some kind of steady state. "We're imagining the universe is evolving its laws by teaching itself," he said. The theory builds on the idea, discussed in chapter 6, that the fundamental physical constants took on their values early in cosmic history, except that now the values are not merely random; there is a logic to them. He and his authors aren't clear on what this logic is, though, so their approach still needs work.

DOES THE BRAIN REQUIRE SPACETIME PHYSICS?

So space emerges in physics, spatial experience emerges in the brain, and the two processes have some suggestive similarities. How far can we push the comparison?

Very far, some think. The philosopher Colin McGinn suggested in the '90s that consciousness involves literally the same physics as did the emergence of space at the dawn of time. Neural processes may create or destroy new dimensions of space in your head. "The brain puts into reverse, as it were, what the big bang initiated," he wrote.[66] This is a much more thoroughgoing appeal to exotic physics than Roger Penrose's ideas from chapter 4. Like many speculative theories about the mind, McGinn's took off from the hard problem

of consciousness. The essence of the hard problem is that consciousness is nonspatial. Our experience is not reducible to parts, as things
in space are, so it is hard to explain by the application of the usual
mechanical laws of physics. So, McGinn proposed, if consciousness
eludes ordinary spatial physics, why not try nonspatial physics? A few
other physicists, most notably David Bohm, have made similar arguments, but never really got beyond the level of metaphor.[67]

Donald Hoffman, a psychologist at the University of California,
Irvine, has gone perhaps the furthest in connecting spatial experience
to emergent space.[68] "All my colleagues in the neurosciences assume
spacetime is fundamental and the contents of spacetime are fundamental and have causal powers," he told me. "If that's not true, then
all our approaches to the hard problem of consciousness have been
founded on an utterly false assumption."

Hoffman starts from a position of philosophical skepticism. Back
in chapter 5, I quoted researchers who doubt that quantum mechanics is a faithful representation of the world. Hoffman doubts that
physics of any sort is a faithful representation. Our brains evolved
for survival, not representational accuracy—they fashion a field of
experience that gets us to do things to stay alive and have lots of kids.
If anything, perceiving the universe as it really is would get in the
way. Hoffman thus puts a biologist's spin on the Kantian view that
basic aspects of our experience, including space and time, reflect the
requirements of thought rather than actual conditions out there in
the world.[69] "It's a rookie mistake," Hoffman said. "We thought we
were seeing reality."

What we *can* be sure of is that we exist. So Hoffman takes
conscious experience as primitive; like Markus Müller in chapter 6,
he is basically a philosophical idealist. "I'm taking experiences to be
absolutely real and foundational, as opposed to space and time and
particles and their properties," he said. His idea is that if we cannot
derive consciousness from physics, then maybe we can derive physics
from consciousness.

Hoffman and Chetan Prakash, an emeritus mathematics professor at Cal State San Bernardino, have performed computer simulations that model the world as a giant social network. The virtual
creatures that inhabit it, which are similar to Hartle's IGUSes, don't
exist in an environment or even in space and time—all they have is
one another. Each constructs a notion of time from the sequence of
interactions it has with its fellows. Through their collective interactions, the creatures reach a mutually consistent standard of time.[70]
Hoffman and Prakash likewise recover space as a convention that
enables social interaction. "Spacetime is a data structure and nothing else," Hoffman said. The mathematics resembles that of some
theories of emergent space. We may think that the universe existed
for billion of years before intelligent beings evolved, but really the
beings came first, he concluded.

Much as I try to be open to new ideas, McGinn's and Hoffman's
are a stretch. Spacetime physics isn't as promising a place to look
for a solution to the hard problem of consciousness as McGinn suggests. Spacetime theorists don't dispense with geometric intuitions
entirely. In their theorizing about emergent space, they rely on abstract spaces, so their theories don't really create a new nonspatial
category of explanation. As for Hoffman and Prakash's ideas, they
strike me as empirically incoherent. If our perceptions of the world
are as far off as they say, how can we trust anything, including the
science that they base their simulations on?

A general problem with supposing that the brain has anything to
do with the physics of emergent space is the yawning gulf in scale. If
space is emergent, then if you probe deep within subatomic particles,
you should eventually see space dissolve into its constituent "atoms."
Traces of this submicroscopic breakdown might potentially manifest
themselves at an observable scale, giving physicists a means to test
these theories, but these traces would be very, very subtle. Nothing
else in physics is so far from human experience; even the vast disparity
between human and cosmic scales pales in comparison. And that is

good. Our very existence depends on the stability of space. We really don't want it falling apart on us.

But Hoffman has gotten a respectful hearing at physics and philosophy meetings I've been to. Maybe people were just being polite, but I do think he makes an important point in saying that reality is too rich for our physics. Physicists know full well that their theories are incomplete, but usually they mean they need to add particles to explain dark matter or reconceive of gravity to explain black holes. When it comes to consciousness, such tweaking won't do. Any idea for expanding our framework to include consciousness is going to sound crazy . . . until it doesn't.

EPILOGUE: IS IT REALLY SO HARD?

THE LATE STRING THEORIST Joe Polchinski told me he loved the title of my first book, *The Complete Idiot's Guide to String Theory*, because the theory made even him feel like an idiot. If a leading physicist felt this way, what hope is there for the rest of us? When someone complains that physics or neuroscience is hard, most people reply, Yeah, tell me something I don't know. But one of the things I learned while researching and writing this book is that many physicists and neuroscientists are just as perplexed as the rest of us.

I routinely hear them worrying that the origins of space or of conscious experience are beyond human comprehension. This is saying something, considering that scientists are disposed to be optimistic—in this line of work, you have to be. The difficulties seem like more than speed bumps; they look disturbingly like the end of the road.

"A cat can't understand calculus," noted Vijay Balasubramanian. He is, as far as I know, the only scientist who works at a serious, nondabbling level on all three topics of this book: fundamental physics, neuroscience, and AI. He sees them as three allies in a single quest. To know how the world works, it's not enough to understand its laws. You also have to understand how we came to know those laws—our own thought processes. "Understanding those limitations or possibilities—the structures of the mind and brain—is, for me, a necessary and complementary part of the quest to write down those fundamental laws. What if the reason why we've struggled to write down a unified theory of all of nature is because the representations the brain is capable of don't permit us to do that?" Balasubramanian said.

We explain things by breaking them down into parts—an essentially spatial mode of reasoning—but this reductionist mindset doesn't always work. Quantum entanglement is puzzling because it is nonspatial. Consciousness, too, seems impervious to spatial analysis. Pessimism about our prospects for solving these problems has a catchy name: mysterianism.[1] Many leading thinkers are mysterians: Colin McGinn, Steven Pinker, Noam Chomsky.[2] They differ from panpsychists in that they think consciousness is a product of nature rather than an exotic add-on. It's just that they think we will never grasp how it works. To resist this dispiriting conclusion requires some fresh thinking.

SCARLET IS LIKE A TRUMPET

Throughout much of this book, I have kept coming back to the ideas of the physicist Carlo Rovelli. He has been arguing for decades that reality consists not of things, but of relations. Einstein's theories, quantum physics, and general considerations about scientific reasoning all indicate to him that there are no observer-independent absolutes. To him, all truths are a matter of perspective. To be clear, "perspective" in this context has a specific mathematical definition; physics is still, by human standards, hard fact.

A relational conception of reality dissolves the hard problem of matter. During the first year of the COVID pandemic, Rovelli pushed this idea further and began to argue that relationalism would also solve the hard problem of consciousness.[3] One way of framing the problem is that third-person physics isn't up to the task of explaining first-person experience and, specifically, its qualitative aspect (qualia). Rovelli gets around that by saying physics is not, in fact, third-person; it is specific to each of us, just as each of us has our own private stream of consciousness; the two sides are not so mismatched after all. In other words, he wants to turn the inside/outside problem into a Möbius strip, having only a single side. So far, though, this is the germ of an idea rather than a fully developed line of inquiry.

A stumbling block for Rovelli's project is that, although physics may well be relational, subjective experience doesn't seem to be. It has an intrinsic *je ne sais quoi* that we grasp without reference to anything else. The sunset is red, period. Its redness may evoke blood or roses, but those are secondary to the immediate experience. To solve the hard problem, Rovelli would need to do one of two things. One would be to do a U-turn and decide that physics is actually *non*relational. He would have to suppose that objects have intrinsic properties on top of the relational properties that the laws of physics capture. Panpsychists propose just that: They think these intrinsic properties are in fact conscious experiences. To them, even something as simple as an electron has some modicum of mind. In 2022 Rovelli, at the urging of the philosopher Emily Adlam, did add nonrelational elements to his interpretation of quantum mechanics.[4] Alternatively, Rovelli could go all-in on relationalism if he somehow showed that the qualities of experience are relational despite feeling intrinsic. Many theories of consciousness have taken a stab at this, including both of those I focus on in this book—integrated information theory (IIT) and predictive coding, introduced in chapter 3.

IIT supposes, like Rovelli, that all objects are ultimately bundles of relations—specifically, causal relations.[5] According to IIT, a subjective experience isn't primitive and unanalyzable; it depends on what the neurons in a neural network are doing now and what they could do next.[6] In principle, you could analyze the network's transitions and read its mind. As I discussed in chapter 8, certain types of network have the feeling of being situated within space; others, of seeing colors. Our own feelings of space and color may arise in this way. "What IIT tries to do is completely avoid any intrinsic quality in the traditional sense," Giulio Tononi said.

Predictive-coding theory reaches the same end by a different route. According to this theory, experiences are predictions we make about the world, and they have a qualitative aspect because we include ourselves in the prediction; qualia are the reasons we use to explain why we react the way we do. When our brain forms a prediction of how we'll respond to these stimuli, it then does what the philosopher Daniel Dennett has called a "strange inversion": it ascribes this prediction not to ourselves, but to the thing we're responding to.[7] Usually we say that when we feel pain, we seek to avoid whatever is causing it; we say that babies are cute, so we coo; we say that honey is sweet, so we crave it. But what may really be going on is that we reflexively seek to escape from the things that harm us, and pain is the story we tell ourselves about why; we attribute cuteness to a baby, when it's really a statement about our own evolved response; we think honey is sweet *because* we crave it.[8] Pain, cuteness, and sweetness seem unanalyzable to us, but by undoing the strange inversion, we can, in fact, analyze them in terms of our own biology, using the standard relational language of science.

A leading advocate of predictive coding, the philosopher Andy Clark at the University of Sussex, put it to me this way: "There's pain because pain is just a simplified way to point to a whole web of dispositions: to move towards or move away from things, to try to avoid those things, take painkillers, all of that stuff. If someone then says, 'Well, why does the pain hurt?' I think what we want to say is, 'Because that's what hurting just *is*. That sense of proclivities to move away from, to take painkillers—all the things we do that are distinctive for pain rather than pleasure.'"

When I began to think about how qualia might be relational, I came across a paper from 2014 by a young philosopher at the University of Helsinki, Kristjan Loorits,[9] and emailed him in 2020; we later met in Helsinki. To my surprise, he had gone off the idea. Efforts to explain qualia relationally are usually told from a third-person perspective, whereas the hard problem of consciousness concerns the first-person perspective. So his current thinking was that qualia may well be relational behind the scenes, but as long as they feel intrinsic to us, they still elude scientific description. "There is still a hard problem in a sense that we seem to be able to experience qualia without being aware of their relational components," Loorits said.

I've been assuming all along that qualia feel intrinsic to us. One way to rescue the idea of relational qualia is to deny that. In 1991 Dennett famously argued that conscious experience is all a big misunderstanding.[10] If someone asks you whether you're conscious, or if you ask yourself, you'll answer, "Of course, silly," but maybe that's just plain wrong, Dennett suggested. After all, when we answer this question, we are reflecting back on having been conscious a moment ago, and this retroactive judgment might be a convenient fiction. "We cannot distinguish an event that we experienced from one that we didn't experience but only remember," the cognitive scientist Joscha Bach told me.

Really? Telling someone they're not conscious smacks of philosophical gaslighting. Even advocates of this "illusionist" approach to consciousness admit they have yet to explain how we could be so badly deluded.[11]

But maybe you don't need to go quite as far as Dennett and Bach do. Loorits suggested a less dismissive position. Maybe qualia are intrinsic *to us*—that is, they are perspectival.[12] That would dovetail with Rovelli's broader program to develop a perspectival physics. We don't ordinarily pay much attention to the nature of qualia. They feel intrinsic only inasmuch as we don't give them further thought. But we could probe deeper. Reflecting on our experience, we might see that what we take to be intrinsic is really relational. For example, Loorits suggested that with artistic training or brain stimulation we could look beneath the intrinsic nature of qualia to see the raw associations that make them up, just as a musician hears the individual components in what, to most fans, is a wall of sound. "It should be possible to experience parts of those underlying structures directly, just as we can learn to experience the individual overtones of a sound," he said. (Loorits knows whereof he speaks: he was a concert pianist before going into philosophy.)

The proposition, then, is that redness, pain, and the other qualities of experience are not Platonic ideals, but a blurred view of a dense thicket of relations. Red is red not because it just is, but because of a vast number of associations that we have learned or been born with. Some have speculated that all our experiences can be placed into a vast "qualia space," in which each quale is defined in relation to every other quale. Qualia might not be as utterly unlike one another as they seem.[13] In recent years, the psychologist Nao Tsuchiya at Monash University and his colleagues have sought to map these relations as a way to test IIT.[14]

If redness is an intrinsic quality of experience, you have to see it to know it,[15] whereas if it is relational, a friend could explain red to you—explain it so fully that, when you finally do see something red, you go, "Yup, just what I thought." To be sure, that explanation would have to go way beyond textbook physics; red isn't just a wavelength of light, but a set of responses within our minds and bodies. Your friend might start by comparing scarlet to the sound of a trumpet.[16] They might also liken the color wheel—in which hues cycle from red to yellow to green to blue to violet to red again—to musical octaves.[17] Through an accumulation of metaphors, they'd communicate to you everything that red means to them, until you achieved the experience of red without ever having seen it for yourself. From the associations in language, psychologists have found, blind people learn the same color relationships as sighted people do.[18]

Helen Keller described understanding sights and sounds by comparison to touch: "Sweet, beautiful vibrations exist for my touch, even though they travel through other substances than air to reach me. So I imagine sweet, delightful sounds, and the artistic arrangement of them which is called music."[19] Even artificial neural networks, which lack not only vision but also any other form of sensory input that could serve as a reference point, can develop a model of color from a purely linguistic analysis.[20]

I had a fascinating chat about the power of metaphor with the theoretical physicist Robbert Dijkgraaf, who was then at the Institute for Advanced Study in Princeton. Dijkgraaf is a synesthete: he sees the numeral 5 as blue, for example. As wondrous as this perceptual ability sounds, he told me it's basically the same as considering a single intellectual problem from different angles. "The idea of having two sensory perceptions of one object is something you see a lot in science," he said. And that is within everyone's ability. Most people apart from artists don't see a need to keep rethinking the concept of "red." But maybe we should. Those of us not born synesthetes can still learn to see red and other primary sensations from multiple perspectives.

If this idea is on the right track, qualia are relational from both the first- and third-person points of view, bridging the explanatory gap. Physics wouldn't need to expand its explanatory repertoire to explain consciousness—we could use math after all. The hard problem would seemingly disappear.

AS LONG AS qualia seem intrinsic to us—either because they really are or because we haven't yet developed the habits of mind that let us see their relational nature—science as we now practice it will be at a loss to describe them. But in 2007 the cognitive scientist Ron Chrisley suggested how science might expand its vocabulary.

Right now, the only acceptable evidence in a conference talk or journal paper is data or mathematical deduction. The messy backstory—a vivid dream, a laboratory screwup—is left to us journalists. "It's considered vulgar to have that kind of phenomenological component as part of one's work," Chrisley said. This taboo is a strength of science; scientific findings leave behind their cultural baggage and stand for the ages. But it deprives researchers of modes of explanation that some topics, including consciousness, may demand.

Chrisley argues that it's perfectly valid for researchers to cite personal experiences in their explanations. These could be experiences of ourselves or of other people, perhaps elicited through art or meditation, but Chrisley focuses on experiences of interacting with our AI creations.[21] In a virtuous cycle he calls "interactive empiricism," engineers instill features of consciousness into their systems, the systems respond, the engineers adapt, and before you know it, they've discovered new principles of consciousness that apply to humans, too. "We are going to be building systems not with the goal of them being conscious, but with them being the kinds of things that prompt the kind of conceptual change in *us* that will eventually allow us to make the next-generation attempt," he said. "You keep spiraling in this way, and at the end we have machine consciousness." When these engineers write up their findings, they won't include just their code and data, but also coaching for evoking the experiences their readers will need to have to follow along. As long as the experiences are reproducible, they're entirely consistent with the other methods of science.

So far, engineers haven't made a concerted effort to build conscious machines, and it may not even be possible, but there have been a few scattered projects over the years, such as the predictive-coding robots created by Jun Tani and colleagues, which have illuminated aspects of human psychology ranging from our perceptual blind spots to conditions such as schizophrenia and autism.[22] Rudimentary though they are, the systems have started to move the deeply felt dividing line between the mechanical and the organic.

WITH A LITTLE HELP FROM OUR AI FRIENDS

In other ways, too, AIs can partner with us to crack the biggest problems in science. They are already analyzing particle collider data, taking imprints of quantum wave functions, predicting material properties, simulating galaxies, and designing lab experiments.[23] "This is an honest-to-god moment in the history of science," said Kyle Cranmer.

Today's machine systems are even able to discover laws of physics. A bundle of techniques known as symbolic regression—like the better-known method of linear regression found in software such as Microsoft Excel, which draws a straight trend line through data points, except not limited to straight lines—can fit an algebraic formula to data.[24] Similar techniques unveil the underlying dynamics producing the data: given the path of a ball, they return the laws of gravity and motion that determined the path. Such systems have rediscovered Kepler's laws of planetary orbits, the laws of electromagnetism, the equations of fluid flow, and principles of nonequilibrium thermodynamics.[25] What took humans centuries to discover, machine systems have recapitulated in a few years.

The hope is that they'll now go out and find laws we never knew of. They could be let loose on archival data, trawling for patterns like a drug company screening thousands of compounds for new drugs. "It would be cool if we could one day discover unknown formulas," said Max Tegmark, who has developed systems with this aim. But it's already starting to happen: in 2019 three theorists used a neural network to discover a new law of knots. Knots are puzzling not just to people untangling their electrical cords, but also to mathematicians and particle theorists. The network found a relation between a knot's twistiness and its so-called hyperbolic volume, a measure of its size.[26] "Nobody had expected that there would be a relation between these numbers and the hyperbolic volume, but the machine learned it," Balasubramanian told me.

As amazing as these machines are, though, when you get into the details, you realize they aren't going to take physicists' jobs anytime soon. They're great at grunt work, which, to be honest, is 90 percent of physics. You might think physicists are sitting around thinking deep thoughts all day, when really they're scratching out pages of work because they accidentally wrote a + as a −. This mechanical labor is best left to a mechanical system. But AI scientists need a lot of hand-holding. For an AI system to extract equations from data, for example, you have to specify the palette of functions that the system will mix and match—sines, cosines, exponentials, and so on—as well as parameters governing the search strategy. What is more, the system's output won't be an answer, but a whole list, leaving the final choice up to you. From what I've seen, you can't get good results out of these systems unless you have a fair amount of experience doing these problems by hand.

These systems are, of course, getting better all the time. Nonetheless, they face fundamental limits. Physics is hard—in a technical, mathematical sense. The number of possible solutions to any problem is vast, and the time it takes to find the right one scales up exponentially with problem size.[27] Even the fastest computer gets only so far, because it has to narrow the search no less than flesh-and-blood physicists do. And the developers of these systems have built in so many prior assumptions that when a machine rediscovers Kepler's laws, it hasn't done anything as impressive as Johannes Kepler himself did.[28] "I think these AI scientists will do the same as human scientists have done, but more and faster," said the AI pioneer Jürgen Schmidhuber of the Dalle Molle Institute for Artificial Intelligence Research in Lugano. "I don't think we are fundamentally worse."

What is emerging, then, is a human-machine partnership. Imagine a counterfactual history in which computers evolved first. They might well invent a new type of device called a "human" to cover their blind spots. This newfangled type of squishy robot, which doesn't require much power but gets by mostly on coffee and occasional praise, does the lateral thinking. Although it tends to make silly mistakes, those imperfections are a benefit, since mistakes are the source of almost all genuine novelty.[29] (Kepler did not produce his namesake laws purely by crunching raw astronomical data; he was inspired by occult ideas about magnetism that were nutty even in his day.)[30]

To unleash the full power of AI, we need to exploit these man-machine differences. It's not enough for machines to do what we've done, only more of it. We need to send them off in some completely new direction. They may not be inherently smarter than we are, but they can be a different kind of smart. "It may well be, for example, that . . . machine intelligence is somehow fundamentally different, and so it's capable of writing things that we're not even capable of doing," Balasubramanian said. He gave an example of one useful difference: machine-learning systems are looser in their reasoning, while physics has a rigid notion of truth that it inherited from mathematics. "In math, we're used to proving theorems saying that if X, then Y," Balasubramanian said. This logic is built into physicists' derivations. They hedge their findings, admitting that an equation holds only under limited circumstances, but this paradigm still closes the door to alternative modes of description. A neural network never makes any such categorical statements; it offers only statistical generalizations, so it might enter domains of theory that the certitude of mathematics closes off. "I think there's some interesting new notion there of the way we should be doing modeling. I think it's a notion of 'probably, approximately correct truths.'"

Another way for AI systems to help is by finding new ways to formulate existing theories, such as quantum mechanics. Physicists already have distinct but equivalent pictures of quantum mechanics developed by Erwin Schrödinger and Werner Heisenberg, each illuminating certain qualities of the quantum world. "Maybe there is a third or fourth representation," suggested the quantum physicist Renato Renner. "If so, our hope is that machine-learning methods may discover it." In one study, he and his colleagues set up several neural networks to work together to describe quantum systems. With humans out of the loop, the networks were free to develop their own private language, like scientists speaking in jargon that is incomprehensible to everyone else.[31]

At a conference I went to in 2016, the computer scientist Bart Selman at Cornell University gave another example of how machines can grasp concepts we can't.[32] He had worked on computer proofs of a mathematical puzzle known as the Erdős discrepancy conjecture. In 2014 a machine filled in an essential step with a proof of 10 billion steps. That sounds long, but it was still much shorter than a brute-force search of all the puzzle solutions; this meant the machine had achieved some genuine understanding. The following year, the mathematician Terence Tao of UCLA published a pithier proof that credited the computer's guidance.[33]

That no human can follow a proof of 10 billion steps is not a failure, but a success. It shows that computers think differently than we do. Computers sometimes see right through problems that stump us, while getting hung up on those we think are easy; entire websites and subreddits are devoted to the silly mistakes that neural networks make. Perhaps the machines will help us the most when they are their most inscrutable.

THE UNIVERSALITY OF HUMAN THOUGHT

But what about Balasubramanian's cat that can't understand calculus? Could we be similarly unable to grasp the origins of the mind or the universe? This is perhaps unknowable, but Dennett noted a huge difference between us and cats: A cat doesn't curl up on a sofa worrying about calculus. It doesn't even know what calculus is. We, on the other hand, obsess over consciousness. We can formulate the hard problem. That alone, according to Dennett, is reason to think we can solve it.[34]

Balasubramanian agrees. He suspects that our thinking processes may have the same universal problem-solving ability that the mathematician Alan Turing proved in the 1930s for computers.[35] "A point of hope is things like the universality of Turing machines—basically, that any computation that can be done can be done by a Turing machine," he said. "In some sense, if the human mind or brain is complete in that sense, maybe it can, through some iterative process, write down complete theories of nature." It's ironic that Chomsky and Pinker doubt we'll ever understand consciousness, because their own academic work has shown that human reasoning is infinitely extensible.[36] We form new concepts without limit by stringing together what we already know or nesting one concept within another. In fact, some suggest that consciousness evolved precisely to allow for open-ended learning.[37]

David Chalmers may be right that consciousness is a different class of problem, and I am not one to make arrogant presumptions about our abilities—our brains are an evolutionary crazy quilt. But there is as yet no sign that science has hit a wall. Our minds evolved to understand the world, which requires that the world be understandable. And we are of this world.

NOTES

1. THE TWIN HARD PROBLEMS

1. Garry Kasparov, *Deep Thinking: Where Machine Intelligence Ends and Human Creativity Begins* (New York: PublicAffairs, 2017), 72–73.
2. David Silver et al., "Mastering the Game of Go with Deep Neural Networks and Tree Search," *Nature* 529, no. 7587 (28 January 2016): 484–89.
3. Ninareh Mehrabi et al., "A Survey on Bias and Fairness in Machine Learning," *ACM Computing Surveys* 54, no. 6 (July 2021): 1–35.
4. Kevin Roose, "Bing's Chatbot Drew Me In and Creeped Me Out," *New York Times*, 17 February 2023.
5. David Kaiser, "When Fields Collide," *Scientific American* 296, no. 6 (June 2007): 62–69.
6. Max Tegmark, "Parallel Universes," *Scientific American* 288, no. 5 (May 2003): 40–51.
7. Freeman J. Dyson, "Time Without End: Physics and Biology in an Open Universe," *Review of Modern Physics* 51, no. 3 (July 1979): 447–60.
8. Philip Goff, *Galileo's Error: Foundations for a New Science of Consciousness* (New York: Pantheon, 2019), 15–19.
9. Margaret A. Boden, *Mind as Machine: A History of Cognitive Science*, 2 vols. (New York: Oxford University Press, 2006), 58–81.
10. William Seager, "Neutral Monism and the Scientific Study of Consciousness" (lecture, Mathematical Consciousness Science, 15 December 2020).
11. Abner Shimony, "Role of the Observer in Quantum Theory," *American Journal of Physics* 31, no. 10 (October 1963): 755–73.
12. John S. Bell, *Speakable and Unspeakable in Quantum Mechanics* (New York: Cambridge University Press, 2004), 117.
13. Brian Greene, *The Hidden Reality: Parallel Universes and the Deep Laws of the Cosmos* (New York: Knopf, 2011), 49–58.

14. Erwin Schrödinger, *Nature and the Greeks* (Cambridge: Cambridge University Press, 1954), 90.

15. Werner Heisenberg, "The Representation of Nature in Contemporary Physics," *Daedalus* 87, no. 3 (Summer 1958): 104–105.

16. John Archibald Wheeler, "Genesis and Observership," in *Foundational Problems in the Special Sciences*, ed. Robert E. Butts and Jaakko Hintikka (Dordrecht: Springer Netherlands, 1977), 27.

17. David J. Chalmers, "Explaining Consciousness Scientifically: Choices and Challenges" (lecture, Science of Consciousness, 12 April 1994).

18. Helen Keller, *The World I Live In* (New York: Century, 1908), 105.

19. Frank Jackson, "Epiphenomenal Qualia," *Philosophical Quarterly* 32, no. 127 (April 1982): 127.

20. Gottfried Leibniz, *Leibniz's Monadology: A New Translation and Guide*, trans. Lloyd Strickland (Edinburgh: Edinburgh University Press, 2014), 17.

21. Liam P. Dempsey, "Thinking-Matter Then and Now: The Evolution of Mind-Body Dualism," *History of Philosophy Quarterly* 26 (January 2009): 43–61.

22. Joseph E. LeDoux, Matthias Michel, and Hakwan Lau, "A Little History Goes a Long Way Toward Understanding Why We Study Consciousness the Way We Do Today," *Proceedings of the National Academy of Sciences of the United States of America* 117, no. 13 (31 March 2020): 6976–84.

23. David J. Chalmers, "Dirty Secrets of Consciousness" (lecture, Foundational Questions Institute Fifth International Conference, Banff, Canada, 18 August 2016).

24. Anil K. Seth, *Being You: A New Science of Consciousness* (New York: Penguin Random House, 2021), 21.

25. Stanislas Dehaene, *Consciousness and the Brain: Deciphering How the Brain Codes Our Thoughts* (New York: Penguin, 2014), 91–92.

26. David J. Chalmers, *The Conscious Mind: In Search of a Fundamental Theory* (New York: Oxford University Press, 1996), 153.

27. Goff, *Galileo's Error*, 122–28.

28. Derk Pereboom, *Consciousness and the Prospects of Physicalism* (New York: Oxford University Press, 2011), 92–100.

29. Joseph Polchinski, *String Theory: An Introduction to the Bosonic String* (New York: Cambridge University Press, 1998), 112–13.

30. Goff, *Galileo's Error*, 175–81.

2. THE NEURAL NETWORK REVOLUTION

1. Alexander Bain, *Mind and Body: The Theories of Their Relation* (New York: D. Appleton, 1875), 109–16; B. Farley and W. Clark, "Simulation of Self-Organizing Systems by Digital Computer," *Transactions of the IRE Professional Group on Information Theory* 4, no. 4 (September 1954): 76–84.

2. Margaret A. Boden, *Mind as Machine: A History of Cognitive Science*, 2 vols. (New York: Oxford University Press, 2006), 911–23.

3. George Musser, "Build Your Own Artificial Neural Network. It's Easy!," *Nautilus* (20 September 2020), https://nautil.us/build-your-own-artificial-neural -network-its-easy-237976/.

4. Brett J. Kagan et al., "*In Vitro* Neurons Learn and Exhibit Sentience When Embodied in a Simulated Game-World," *Neuron* 110, no. 23 (7 December 2022): 3952–69.e8.

5. Boden, *Mind as Machine*, 79–80, 123–28.

6. Howard Crosby Warren, *A History of the Association Psychology* (New York: Charles Scribner's Sons, 1921), 23–28.

7. David Hartley, *Observations on Man, His Frame, His Duty, and His Expectations* (London: T. Tegg and Sons, 1834).

8. John Sutton, *Philosophy and Memory Traces* (New York: Cambridge University Press, 1998), chaps. 6 and 13.

9. David Cahan, *Helmholtz: A Life in Science* (Chicago: University of Chicago Press, 2018), 61–63.

10. Kate Harper, "Alexander Bain's *Mind and Body* (1872): An Underappreciated Contribution to Early Neuropsychology," *Journal of the History of the Behavioral Sciences* 55, no. 2 (April 2019): 139–60.

11. Bain, *Mind and Body*, 109–16; William James, *The Principles of Psychology* (New York: Holt, 1890), 566–70.

12. Bain, *Mind and Body*, 119n3.

13. D. O. Hebb, *The Organization of Behavior: A Neuropsychological Theory* (New York: Wiley, 1949), 62.

14. Bain, *Mind and Body*, 44–50.

15. Stephen Grossberg, "How Does a Brain Build a Cognitive Code?," *Psychological Review* 87, no. 1 (January 1980): 1–51.

16. Steve J. Heims, *The Cybernetics Group* (Cambridge, MA: MIT Press, 1991), 11–12, 34–35.

17. Boden, *Mind as Machine*, 903–11.

18. Heims, *Cybernetics Group*, 95–96.

19. Michel Morange, *The Black Box of Biology: A History of the Molecular Revolution* (Cambridge, MA: Harvard University Press, 2020), ch. 7.

20. Judith P. Swazey, "Forging a Neuroscience Community: A Brief History of the Neurosciences Research Program," in *The Neurosciences: Paths of Discovery*, ed. George Adelman et al. (Cambridge, MA: MIT Press, 1975), 529–46.

21. John J. Hopfield, "Neural Networks and Physical Systems with Emergent Collective Computational Abilities," *Proceedings of the National Academy of Sciences* 79, no. 8 (15 April 1982): 2554–58.

22. Trenton Bricken and Cengiz Pehlevan, "Attention Approximates Sparse Distributed Memory" (preprint, submitted 10 November 2021).

23. John J. Hopfield and David W. Tank, "'Neural' Computation of Decisions in Optimization Problems," *Biological Cybernetics* 52, no. 3 (1985): 141–52.

24. John J. Hopfield, "Searching for Memories, Sudoku, Implicit Check Bits, and the Iterative Use of Not-Always-Correct Rapid Neural Computation," *Neural Computation* 20, no. 5 (May 2008): 1119–64.

25. W. A. Little, "The Existence of Persistent States in the Brain," *Mathematical Biosciences* 19, no. 1–2 (February 1974): 101–20; Shun-Ichi Amari, "Neural Theory of Association and Concept-Formation," *Biological Cybernetics* 26, no. 3 (17 May 1977): 175–85.

26. Douglas R. Hofstadter, "Waking Up from the Boolean Dream, or, Subcognition as Computation," in *Metamagical Themas: Questing for the Essence of Mind and Pattern* (New York: Basic Books, 1985), 631–65.

27. Boden, *Mind as Machine*, 936–46.

28. B. P. Abbott et al., "Observation of Gravitational Waves from a Binary Black Hole Merger," *Physical Review Letters* 116, no. 6 (12 February 2016): 061102.

29. Dana H. Ballard, Geoffrey E. Hinton, and Terrence J. Sejnowski, "Parallel Visual Computation," *Nature* 306, no. 5938 (3 November 1983): 21–26.

30. Geoffrey E. Hinton and Terrence J. Sejnowski, "Optimal Perceptual Inference," *Proceedings of the IEEE Conference on Computer Vision and Pattern Recognition* 448 (June 1983): 448–53.

31. Kim Sharp and Franz Matschinsky, "Translation of Ludwig Boltzmann's Paper 'On the Relationship Between the Second Fundamental Theorem of the Mechanical Theory of Heat and Probability Calculations Regarding the Conditions for Thermal Equilibrium,'" *Entropy* 17, no. 4 (April 2015): 1971–2009.

32. John von Neumann, "The General and Logical Theory of Automata," in *Cerebral Mechanisms in Behavior: The Hixon Symposium*, ed. Lloyd A. Jeffress (New York: Hafner, 1967), 1–42.

33. David H. Ackley, Geoffrey E. Hinton, and Terrence J. Sejnowski, "A Learning Algorithm for Boltzmann Machines," *Cognitive Science* 9, no. 1 (January 1985): 147–69.

34. Ian J. Goodfellow et al., "Generative Adversarial Networks" (preprint, submitted 10 June 2014).

35. David E. Rumelhart, Geoffrey E. Hinton, and Ronald J. Williams, "Learning Representations by Back-Propagating Errors," *Nature* 323, no. 6088 (9 October 1986): 533–36.

36. Yann LeCun, "A Theoretical Framework for Back-Propagation," in *Proceedings of the 1988 Connectionist Models Summer School*, ed. David S. Touretzky, Geoffrey E. Hinton, and Terrence J. Sejnowski (San Mateo, CA: Morgan Kaufmann, 1988), 21–28.

37. James A. Anderson and Edward Rosenfeld, "Geoffrey E. Hinton," in *Talking Nets: An Oral History of Neural Networks* (Cambridge, MA: MIT Press, 2000), 379.

38. Ali Rahimi, "Back When We Were Young" (lecture, Conference on Neural Information Processing Systems, Long Beach, CA, 5 December 2017).

39. Lawrence M. Principe, "Reflections on Newton's Alchemy in Light of the New Historiography of Alchemy," in *Newton and Newtonianism*, ed. James E. Force and Sarah Hutton (Boston: Kluwer Academic, 2004), 2015–19.

40. Chris Olah, Alexander Mordvintsev, and Ludwig Schubert, "Feature Visualization," *Distill*, 7 November 2017, https://distill.pub/2017/feature-visualization/.

41. Michael Hahn and Marco Baroni, "Tabula Nearly Rasa: Probing the Linguistic Knowledge of Character-Level Neural Language Models Trained on Unsegmented Text" (preprint, submitted 17 June 2019).

42. Yann LeCun et al., "Gradient-Based Learning Applied to Document Recognition," *Proceedings of the IEEE* 86, no. 11 (November 1998): 2278–324.

43. Ashish Vaswani et al., "Attention Is All You Need" (preprint, submitted 12 June 2017).

44. Geoffrey E. Hinton, "Connectionist Learning Procedures," *Artificial Intelligence* 40, no. 1–3 (September 1989): 185–234.

45. Terrence J. Sejnowski and Charles R. Rosenberg, "Parallel Networks That Learn to Pronounce English Text," *Complex Systems* 1, no. 1 (February 1987): 145–68.

46. Chiyuan Zhang et al., "Understanding Deep Learning Requires Rethinking Generalization" (preprint, submitted 10 November 2016); Mikhail Belkin et al., "Reconciling Modern Machine-Learning Practice and the Classical Bias-Variance Trade-Off," *Proceedings of the National Academy of Sciences of the United States of America* 116, no. 32 (6 August 2019): 15849–54.

47. Mario Geiger et al., "Jamming Transition as a Paradigm to Understand the Loss Landscape of Deep Neural Networks," *Physical Review E* 100, no. 1 (11 July 2019): 012115.

48. Terrence J. Sejnowski, "The Unreasonable Effectiveness of Deep Learning in Artificial Intelligence," *Proceedings of the National Academy of Sciences* 117, no. 48 (1 December 2020): 30033–38.

49. Surya Ganguli and Haim Sompolinsky, "Compressed Sensing, Sparsity, and Dimensionality in Neuronal Information Processing and Data Analysis," *Annual Review of Neuroscience* 35, no. 1 (July 2012): 485–508.

50. Daniel J. Amit, Hanoch Gutfreund, and Haim Sompolinsky, "Spin-glass Models of Neural Networks," *Physical Review A* 32, no. 2 (August 1985): 1007–18; Radford M. Neal, *Bayesian Learning for Neural Networks: Lecture Notes in Statistics* (New York: Springer New York, 1996), chap 2.

51. Jaehoon Lee et al., "Deep Neural Networks as Gaussian Processes" (preprint, submitted 31 October 2017).

52. Carl Edward Rasmussen, "Gaussian Processes in Machine Learning," in *Advanced Lectures on Machine Learning: Lecture Notes in Computer Science*, ed. Olivier Bousquet, Ulrike von Luxburg, and Gunnar Rätsch (Berlin, Heidelberg: Springer Berlin Heidelberg, 2004).

53. Samuel S. Schoenholz et al., "Deep Information Propagation" (preprint, submitted 4 November 2016).

54. Thierry Mora and William Bialek, "Are Biological Systems Poised at Criticality?," *Journal of Statistical Physics* 144, no. 2 (2 June 2011): 268–302.

55. Sho Yaida, "Non-Gaussian Processes and Neural Networks at Finite Widths" (preprint, submitted 30 September 2019).

56. Richard P. Feynman, "Simulating Physics with Computers," *International Journal of Theoretical Physics* 21, no. 6/7 (1982): 467–88.

57. Elizabeth C. Behrman et al., "A Quantum Dot Neural Network," in *Proceedings of the Fourth Workshop on Physics of Computation*, ed. Tommaso Toffoli, Michael Biafore, and João Leão (Cambridge, MA: New England Complex Systems Institute, 1996), 22–24.

58. Hidetoshi Nishimori and Yoshihiko Nonomura, "Quantum Effects in Neural Networks," *Journal of the Physical Society of Japan* 65, no. 12 (15 December 1996): 3780–96.

59. Tadashi Kadowaki and Hidetoshi Nishimori, "Quantum Annealing in the Transverse Ising Model," *Physical Review E* 58, no. 5 (1 November 1998): 5355–63.

60. M. W. Johnson et al., "Quantum Annealing with Manufactured Spins," *Nature* 473, no. 7346 (12 May 2011): 194–98.

61. Quinten Hardy, "A Strange Computer Promises Great Speed," *New York Times*, 22 March 2013.

62. Jacob Biamonte et al., "Quantum Machine Learning," *Nature* 549, no. 7671 (13 September 2017): 195–202.

63. Hartmut Neven, "Car Detector Trained with the Quantum Adiabatic Algorithm" (demonstration, Conference on Neural Information Processing Systems, Vancouver, Canada, 8 December 2009).

64. Alex Mott et al., "Solving a Higgs Optimization Problem with Quantum Annealing for Machine Learning," *Nature* 550, no. 7676 (19 October 2017): 375–79.

65. Vasil S. Denchev et al., "What Is the Computational Value of Finite-Range Tunneling?," *Physical Review X* 6, no. 3 (1 August 2016): 031015.

66. Maria Schuld, Ilya Sinayskiy, and Francesco Petruccione, "Quantum Computing for Pattern Classification," in *Lecture Notes in Computer Science: PRICAI 2014: Trends in Artificial Intelligence*, ed. Duc-Nghia Pham and Seong-Bae Park (Cham, Switzerland: Springer International, 2014), 208–20.

67. Elizabeth C. Behrman and James E. Steck, "Multiqubit Entanglement of a General Input State," *Quantum Information and Computation* 13, no. 1/2 (2013): 36–53; Edward Farhi and Hartmut Neven, "Classification with Quantum Networks on Near Term Processors" (preprint, submitted 18 December 2017); A. V. Uvarov, A. S. Kardashin, and Jacob D. Biamonte, "Machine Learning Phase Transitions with a Quantum Processor," *Physical Review A* 102, no. 1 (15 July 2020): 012415.

68. Gia Dvali, "Black Holes as Brains: Neural Networks with Area Law Entropy," *Fortschritt der Physik* 66, no. 4 (27 March 2018): 1800007.

69. C. F. von Weizsäcker, "Physics and Philosophy," in *The Physicist's Conception of Nature*, ed. Jagdish Mehra (Dordrecht: Springer Netherlands, 1973), 737.

3. PHYSICS OF THE MIND

1. Jakob Hohwy, *The Predictive Mind* (New York: Oxford University Press, 2013), 5.

2. Piotr Litwin and Marcin Miłkowski, "Unification by Fiat: Arrested Development of Predictive Processing," *Cognitive Science* 44, no. 7 (July 2020): e12867.

3. Johannes Kleiner and Erik P. Hoel, "Falsification and Consciousness," *Neuroscience of Consciousness* 2021, no. 1 (2021): nniab001.

4. David Cahan, *Helmholtz: A Life in Science* (Chicago: University of Chicago Press, 2018), 59.

5. Cahan, *Helmholtz*, 66–70.

6. Cahan, *Helmholtz*, 90–95, 327–30.

7. David M. Eagleman, "How Does the Timing of Neural Signals Map onto the Timing of Perception?," in *Space and Time in Perception and Action*, ed. Romi Nijhawan and Beena Khurana (Cambridge: Cambridge University Press, 2010): 216–31.

8. Hermann von Helmholtz, *Treatise on Physiological Optics,* trans. James P. C. Southall (Rochester, NY: Optical Society of America, 1925), sec. 26.

9. Helmholtz, *Treatise on Physiological Optics*, sec. 24.

10. Helmholtz, *Treatise on Physiological Optics*, sec. 32.

11. Hohwy, *Predictive Mind*, 217.

12. Karl J. Friston, "Hallucinations and Perceptual Inference," *Behavioral and Brain Sciences* 28, no. 6 (22 December 2005): 764–66.

13. Cahan, *Helmholtz*, 310, 327–40.

14. Donald M. MacKay, "The Epistemological Problem for Automata," in *Automata Studies*, ed. Claude E. Shannon and J. McCarthy (Princeton, NJ: Princeton University Press, 1956), 235–52.

15. Donald M. MacKay, "Mindlike Behaviour in Artefacts," *British Journal for the Philosophy of Science* 2, no. 6 (October 1951): 105–21.

16. M. V. Srinivasan, S. B. Laughlin, and A. Dubs, "Predictive Coding: A Fresh View of Inhibition in the Retina," *Proceedings of the Royal Society B: Biological Sciences* 216, no. 1205 (22 November 1982): 427–59.

17. Khalid Sayood, *Introduction to Data Compression*, 4th ed. (Waltham, MA: Morgan Kaufmann, 2012), chaps. 3, 7.

18. Peter Dayan et al., "The Helmholtz Machine," *Neural Computation* 7, no. 5 (September 1995): 889–904.

19. Rajesh P. N. Rao and Dana H. Ballard, "Predictive Coding in the Visual Cortex: A Functional Interpretation of Some Extra-Classical Receptive-Field Effects," *Nature Neuroscience* 2, no. 1 (January 1999): 79–87.

20. Benjamin Kuipers et al., "Shakey: From Conception to History," *AI Magazine* 38, no. 1 (Spring 2017): 88–103.

21. Jun Tani, "Model-Based Learning for Mobile Robot Navigation from the Dynamical Systems Perspective," *IEEE Transactions on Systems, Man, and Cybernetics, Part B (Cybernetics)* 26, no. 3 (June 1996): 421–36.

22. Jun Tani, "Learning to Generate Articulated Behavior Through the Bottom-Up and the Top-Down Interaction Processes," *Neural Networks* 16, no. 1 (January 2003): 11–23.

23. Jakob Hohwy, "Priors in Perception: Top-Down Modulation, Bayesian Perceptual Learning Rate, and Prediction Error Minimization," *Consciousness and Cognition* 47 (January 2017): 75–85.

24. Julian Kiverstein, Mark Miller, and Erik Rietveld, "The Feeling of Grip: Novelty, Error Dynamics, and the Predictive Brain," *Synthese* 196, no. 7 (23 October 2017): 2847–69.

25. George Musser, "How Autism May Stem from Problems with Prediction," *Spectrum News* (7 March 2018), https://www.spectrumnews.org/features/deep-dive/autism-may-stem-problems-prediction/.

26. Karl J. Friston, "Learning and Inference in the Brain," *Neural Networks* 16, no. 9 (November 2003): 1325–52.

27. William James, *The Principles of Psychology* (New York: Holt, 1890), chap. 26.

28. Sam Schramski, "Running Is Always Blind," *Nautilus* 38 (30 June 2016), https://nautil.us/running-is-always-blind-236003/.

29. Karl J. Friston et al., "Dopamine, Affordance and Active Inference," *PLoS Computational Biology* 8, no. 1 (January 2012): e1002327.

30. Karl J. Friston. "I Am Therefore I Think" (lecture, Foundational Questions Institute Sixth International Conference, Castelvecchio Pascoli, Italy, 23 July 2019).

31. Michael D. Kirchhoff and Tom Froese, "Where There Is Life There Is Mind: In Support of a Strong Life-Mind Continuity Thesis," *Entropy* 19, no. 4 (14 April 2017): 169.

32. James Kasting, *How to Find a Habitable Planet* (Princeton, NJ: Princeton University Press, 2012), 49–56.

33. Sergio Rubin et al., "Future Climates: Markov Blankets and Active Inference in the Biosphere," *Journal of the Royal Society Interface* 17, no. 172 (November 2020): 20200503.

34. Artemy Kolchinsky and David H. Wolpert, "Semantic Information, Autonomous Agency and Non-Equilibrium Statistical Physics," *Interface Focus* 8, no. 6 (6 December 2018): 20180041.

35. Jun Tani, "An Interpretation of the 'Self' from the Dynamical Systems Perspective: A Constructivist Approach," *Journal of Consciousness Studies* 5 (1 May 1998): 516–42.

36. Kelsey Klotz, "The Art of the Mistake," *The Common Reader* 11 (Summer 2019), https://commonreader.wustl.edu/c/the-art-of-the-mistake/.

37. René Descartes, *Meditations on First Philosophy*, trans. Michael Moriarty (New York: Oxford University Press, 2008), 60–61.

38. Tim Bayne, "On the Axiomatic Foundations of the Integrated Information Theory of Consciousness," *Neuroscience of Consciousness* 2018, no. 1 (29 June 2018): 159.

39. Pedro A. M. Mediano et al., "Integrated Information as a Common Signature of Dynamical and Information-Processing Complexity," *Chaos* 32, no. 1 (January 2022): 013115.

40. Hyoungkyu Kim and UnCheol Lee, "Criticality as a Determinant of Integrated Information Φ in Human Brain Networks," *Entropy* 21, no. 10 (October 2019): 981.

41. Max Tegmark, "Improved Measures of Integrated Information," *PLoS Computational Biology* 12, no. 11 (21 November 2016): e1005123.

42. Tim Bayne, Jakob Hohwy, and Adrian M. Owen, "Are There Levels of Consciousness?," *Trends in Cognitive Sciences* 20, no. 6 (June 2016): 405–13.

43. Mark S. George, "Stimulating the Brain," *Scientific American* 289 (September 2003): 66–73.

44. Adenauer G. Casali et al., "A Theoretically Based Index of Consciousness Independent of Sensory Processing and Behavior," *Science Translational Medicine* 5, no. 198 (14 August 2013): 198ra105.

45. A. Arena et al., "General Anesthesia Disrupts Complex Cortical Dynamics in Response to Intracranial Electrical Stimulation in Rats," *eNeuro* 8, no. 4 (5 August 2021): ENEURO.0343–20.2021; Roberto N. Muñoz et al., "General Anesthesia Reduces Complexity and Temporal Asymmetry of the Informational Structures Derived from Neural Recordings in *Drosophila*," *Physical Review Research* 2 (22 May 2020): 023219.

46. David Balduzzi and Giulio Tononi, "Qualia: The Geometry of Integrated Information," *PLoS Computational Biology* 5, no. 8 (August 2009): e1000462.

47. Mélanie Boly, "Are the Neural Correlates of Consciousness (Mostly) in the Front or in the Back of the Cerebral Cortex?" (lecture, Association of the Scientific Study of Consciousness, Kraków, Poland, 18 June 2018).

48. Ryota Kanai, "Consciousness and A.I." (lecture, Human-Level AI 2018, Prague, 25 August 2018); Jun Kitazono, Ryota Kanai, and Masafumi Oizumi, "Efficient Search for Informational Cores in Complex Systems: Application to Brain Networks," *Neural Networks* 132 (December 2020): 232–44.

49. Brian Odegaard, Robert T. Knight, and Hakwan Lau, "Should a Few Null Findings Falsify Prefrontal Theories of Conscious Perception?," *Journal of Neuroscience* 37, no. 40 (4 October 2017): 9593–602.

50. Kirchhoff and Froese, "Where There Is Life"; Philip Goff, *Galileo's Error: Foundations for a New Science of Consciousness* (New York: Pantheon, 2019), 138–39.

51. Goff, *Galileo's Error*, 164–69.

52. Karl J. Friston, Wanja Wiese, and J. Allan Hobson, "Sentience and the Origins of Consciousness: From Cartesian Duality to Markovian Monism," *Entropy* 22, no. 5 (May 2020): 17–18.

53. Giulio Tononi and Christof Koch, "Consciousness: Here, There and Everywhere?," *Philosophical Transactions of the Royal Society B: Biological Sciences* 370, no. 1668 (19 May 2015): 13.

54. Giulio Tononi et al., "Integrated Information Theory: From Consciousness to Its Physical Substrate," *Nature Reviews Neuroscience* 17, no. 7 (July 2016): 455.

55. Tim Bayne, Anik K. Seth, and Marcello Massimini, "Are There Islands of Awareness?," *Trends in Neurosciences* 43, no. 1 (January 2020): 6–16.

56. Hedda Hassel Mørch, "Is the Integrated Information Theory of Consciousness Compatible with Russellian Panpsychism?," *Erkenntnis* 84, no. 5 (10 April 2018): 1065–85.

57. Daniel A. Friedman and Eirik Søvik, "The Ant Colony as a Test for Scientific Theories of Consciousness," *Synthese* 198, no. 2 (12 February 2019): 1457–80.
58. Christian List, "What Is It Like to Be a Group Agent?," *Noûs* 52, no. 2 (28 July 2016): 295–319.
59. Margaret A. Boden, *Mind as Machine: A History of Cognitive Science*, 2 vols. (New York: Oxford University Press, 2006), 1356–62.
60. David J. Chalmers, *The Conscious Mind: In Search of a Fundamental Theory* (New York: Oxford University Press, 1996), 84–88.
61. Larissa Albantakis et al., "Evolution of Integrated Causal Structures in Animats Exposed to Environments of Increasing Complexity," *PLoS Computational Biology* 10, no. 12 (18 December 2014): e1003966.
62. Tononi and Koch, "Consciousness," 15.
63. Wanja Wiese and Karl J. Friston, "The Neural Correlates of Consciousness Under the Free Energy Principle: From Computational Correlates to Computational Explanation," *Philosophy and the Mind Sciences* 2 (22 September 2021).
64. Katherine Elkins and Jon Chun, "Can GPT-3 Pass a Writer's Turing Test?," *Journal of Cultural Analytics* 5, no. 2 (14 September 2020).
65. David J. Chalmers, "Could a Large Language Model Be Conscious?" (lecture, Conference on Neural Information Processing Systems, New Orleans, 28 November 2022).
66. Susan Schneider, *Artificial You* (Princeton, NJ: Princeton University Press, 2019), 36.

4. THE QUANTUM BRAIN

1. Roger Penrose, *The Emperor's New Mind: Concerning Computers, Minds, and the Laws of Physics* (New York: Penguin Books, 1991).
2. Penrose, *Emperor's New Mind*, 402–404, 431–33.
3. Roger Penrose, *Shadows of the Mind: A Search for the Missing Science of Consciousness* (New York: Oxford University Press, 1994), 351–52.
4. J. N. Tinsley, et al., "Direct Detection of a Single Photon by Humans," *Nature Communications* 7 (19 July 2016): 12172.
5. John von Neumann, *Mathematical Foundations of Quantum Mechanics*, trans. Robert T. Beyer (Princeton, NJ: Princeton University Press, 1955), chap. 5.
6. Mary B. Hesse, *Forces and Fields: The Concept of Action at a Distance in the History of Physics* (Mineola, NY: Dover, 2005), 90–95.
7. Hugh Everett, "The Theory of the Universal Wave Function," in *The Many Worlds Interpretation of Quantum Mechanics*, ed. Bryce S. DeWitt and Neill Graham (Princeton, NJ: Princeton University Press, 1973), 61.
8. Eugene Paul Wigner, "Remarks on the Mind-Body Question," in *The Scientist Speculates: An Anthology of Partly-Baked Ideas*, ed. Irving John Good (London: Heinemann, 1962), 256.
9. Penrose, *Emperor's New Mind*, 297–98.

10. Sandro Donadi et al., "Underground Test of Gravity-Related Wave Function Collapse," *Nature Physics* 17, no. 1 (January 2021): 74–78.

11. Yaakov Y. Fein et al., "Quantum Superposition of Molecules Beyond 25 kDa," *Nature Physics* 15, no. 12 (23 September 2019): 1242–45.

12. A. D. O'Connell et al., "Quantum Ground State and Single-Phonon Control of a Mechanical Resonator," *Nature* 464, no. 7289 (1 April 2010): 697–703.

13. C. Marletto et al., "Entanglement Between Living Bacteria and Quantized Light Witnessed by Rabi Splitting," *Journal of Physics Communications* 2, no. 10 (10 October 2018): 101001.

14. K. S. Lee et al., "Entanglement in a Qubit-qubit-tardigrade System," *New Journal of Physics* 24, no. 12 (December 2022): 123024.

15. John S. Bell, *Speakable and Unspeakable in Quantum Mechanics* (New York: Cambridge University Press, 2004), 1–21.

16. Radek Lapkiewicz et al., "Experimental Non-Classicality of an Indivisible Quantum System," *Nature* 474, no. 7352 (23 June 2011): 490–93.

17. Tim Maudlin, *Quantum Non-Locality and Relativity: Metaphysical Intimations of Modern Physics*, 2nd ed. (Malden, MA: Blackwell, 2002), 116–21.

18. E. Joos and H. D. Zeh, "The Emergence of Classical Properties Through Interaction with the Environment," *Zeitschrift für Physik B Condensed Matter* 59, no. 2 (June 1985): 242.

19. Maximilian Schlosshauer and Arthur Fine, "Decoherence and the Foundations of Quantum Mechanics," in *The Frontiers Collection: Quantum Mechanics at the Crossroads*, ed. James Evans and Alan S. Thorndike (Heidelberg: Springer Berlin Heidelberg, 2007), 143.

20. Eugene Paul Wigner, "Review of the Quantum Mechanical Measurement Problem," in *Quantum Optics, Experimental Gravity, and Measurement Theory*, ed. Pierre Meystre and Marlan O. Scully (Boston: Springer U.S., 1983), 58; Leslie E. Ballentine, "A Meeting with Wigner," *Foundations of Physics* 49, no. 8 (August 2019): 783–85.

21. Fritz London and Edmond Bauer, "The Theory of Observation in Quantum Mechanics," in *Quantum Theory and Measurement*, ed. John Archibald Wheeler and Wojciech Hubert Zurek (Princeton, NJ: Princeton University Press, 1983), 251–52.

22. Kostas Gavroglu, *Fritz London: A Scientific Biography* (New York: Cambridge University Press, 1995), 169–75.

23. Karl K. Darrow, "Edmond Bauer," *Physics Today* 17, no. 6 (June 1964): 86–87; David Schoenbrun, *Soldiers of the Night: The Story of the French Resistance* (New York: Dutton, 1980), 243–45.

24. David J. Chalmers and Kelvin J. McQueen, "Consciousness and the Collapse of the Wave Function," in *Consciousness and Quantum Mechanics*, ed. Shan Gao (New York: Oxford University Press, 2022), 11–63.

25. Kobi Kremnizer and André Ranchin, "Integrated Information-Induced Quantum Collapse," *Foundations of Physics* 45, no. 8 (19 May 2015): 889–99; Elias

Okon and Miguel Ángel Sebastián, "A Consciousness-Based Quantum Objective Collapse Model," *Synthese* 197, no. 9 (27 July 2018): 3947–67.

26. Max Tegmark, "Consciousness as a State of Matter," *Chaos, Solitons & Fractals* 76 (July 2015): 238–70.

27. Heinrich Päs, *The One: How an Ancient Idea Holds the Future of Physics* (New York: Basic Books, 2023), 251.

28. Penrose, *Emperor's New Mind*, 349.

29. Richard P. Feynman, *Feynman Lectures on Gravitation*, ed. William G. Wagner and Fernando B. Morninigo (New York: Addison-Wesley, 1995), 11–15; F. Károlyházy, "Gravitation and Quantum Mechanics of Macroscopic Objects," *Nuovo Cimento A* 42, no. 2 (March 1966): 390–402; Roger Penrose, "Gravity and State Vector Reduction," in *Quantum Concepts in Space and Time*, ed. Roger Penrose and Christopher J. Isham (New York: Oxford University Press, 1986), 129–46.

30. J. B. Olmsted and G. G. Borisy, "Microtubules," *Annual Review of Biochemistry* 42 (1973): 507–40.

31. Jarema J. Malicki and Colin A. Johnson, "The Cilium: Cellular Antenna and Central Processing Unit," *Trends in Cell Biology* 27, no. 2 (February 2017): 126–40.

32. Stuart R. Hameroff and Richard C. Watt, "Information Processing in Microtubules," *Journal of Theoretical Biology* 98, no. 4 (October 1982): 549–61.

33. Stuart R. Hameroff, *Ultimate Computing: Biomolecular Consciousness and Nano-Technology* (Amsterdam: Elsevier Science, 1987), 14.

34. David Beniaguev, Idan Segev, and Michael London, "Single Cortical Neurons as Deep Artificial Neural Networks," *Neuron* 109, no. 17 (1 September 2021): 2727–39.e3.

35. Ida V. Lundholm et al., "Terahertz Radiation Induces Non-Thermal Structural Changes Associated with Fröhlich Condensation in a Protein Crystal," *Structural Dynamics* 2, no. 5 (September 2015): 054702.

36. Hameroff, *Ultimate Computing*, 197.

37. H. Fröhlich, "Long-Range Coherence and Energy Storage in Biological Systems," *International Journal of Quantum Chemistry* 2, no. 5 (September 1968): 641–49.

38. J. C. Eccles, "Do Mental Events Cause Neural Events Analogously to the Probability Fields of Quantum Mechanics?," *Proceedings of the Royal Society B: Biological Sciences* 227, no. 1249 (22 May 1986): 411–28; Roger J. Faber, *Clockwork Garden: On the Mechanistic Reduction of Living Things* (Amherst: University of Massachusetts Press, 1986); I. N. Marshall, "Consciousness and Bose-Einstein Condensates," *New Ideas in Psychology* 7, no. 1 (1989): 73–83; Henry P. Stapp, "Quantum Propensities and the Brain-Mind Connection," *Foundations of Physics* 21, no. 12 (November 1991): 1451–77.

39. Stuart R. Hameroff, "The Quantum Origin of Life: How the Brain Evolved to Feel Good," in *On Human Nature: Biology, Psychology, Ethics, Politics, and Religion*, ed. Michel Tibayrenc and Francisco J. Ayala (New York: Elsevier, 2017), 334–35.

40. David J. Chalmers, "Consciousness and Its Place in Nature," in *The Blackwell Guide to Philosophy of Mind* (Malden, MA: Blackwell, 2007), 125–27.

41. David J. Chalmers, "Panpsychism and Panprotopsychism," in *Consciousness in the Physical World: Perspectives on Russellian Monism*, ed. Torin Alter and Yujin Nagasawa (New York: Oxford University Press, 2015), 246–76.
42. Stuart R. Hameroff and Roger Penrose, "Consciousness in the Universe: An Updated Review of the 'Orch OR' Theory," in *Biophysics of Consciousness: A Foundational Approach*, ed. Roman R. Poznanski, Jack A. Tuszyński, and Todd E. Feinberg (Singapore: World Scientific, 2016), 517–99.
43. G. C. Ghirardi, A. Rimini, and T. Weber, "Unified Dynamics for Microscopic and Macroscopic Systems," *Physical Review D* 34, no. 2 (15 July 1986): 470–91.
44. Hameroff, "Quantum Origin of Life."
45. Max Tegmark, "Importance of Quantum Decoherence in Brain Processes," *Physical Review E* 61, no. 4 (April 2000): 4194–206.
46. Penrose, *Emperor's New Mind*, 402; Penrose, *Shadows of the Mind*, 351–52.
47. Scott Hagan, Stuart R. Hameroff, and Jack A. Tuszyński, "Quantum Computation in Brain Microtubules: Decoherence and Biological Feasibility," *Physical Review E* 65, no. 6 pt. 1 (June 2002): 061901.
48. Hameroff, "Quantum Origin of Life," 337–43.
49. Huping Hu and Maoxin Wu, "Spin-Mediated Consciousness Theory: Possible Roles of Neural Membrane Nuclear Spin Ensembles and Paramagnetic Oxygen," *Medical Hypotheses* 63, no. 4 (2004): 633–46.
50. Johnjoe McFadden and Jim Al-Khalili, *Life on the Edge: The Coming of Age of Quantum Biology* (New York: Crown, 2014), 119–31.
51. Hugo Cable and Kavan Modi, "Harness Quantum Noise to Unlock Quantum Computing," *New Scientist* 220, no. 2943 (16 November 2013): 30–31.
52. McFadden and Al-Khalili, *Life on the Edge*, 188–95.
53. Jordan Smith et al., "Radical Pairs May Play a Role in Xenon-Induced General Anesthesia," *Scientific Reports* 11, no. 1 (18 March 2021): 6287.
54. McFadden and Al-Khalili, *Life on the Edge*, 155–65.
55. Elizabeth C. Behrman et al., "Microtubules as a Quantum Hopfield Network," in *The Emerging Physics of Consciousness*, ed. Jack Tuszyński (Heidelberg: Springer Berlin Heidelberg, 2006), 351–70.

5. PHYSICS IN THE FIRST PERSON

1. Eugene Paul Wigner, "Remarks on the Mind-Body Question," in *The Scientist Speculates: An Anthology of Partly-Baked Ideas*, ed. Irving John Good (London: Heinemann, 1962), 284–302; Hugh Everett, "The Theory of the Universal Wave Function," in *The Many Worlds Interpretation of Quantum Mechanics*, ed. Bryce S. DeWitt and Neill Graham (Princeton, NJ: Princeton University Press, 1973), 4–8; Jeffrey A. Barrett and Peter Byrne, eds., *The Everett Interpretation of Quantum Mechanics* (Princeton, NJ: Princeton University Press, 2012), 14–15.
2. George Musser, "Schrödinger's A.I. Could Test the Foundations of Reality," *FQxI Blogs*, 19 September 2022, https://fqxi.org/community/articles/display/266.

3. David Deutsch, "Quantum Theory as a Universal Physical Theory," *International Journal of Theoretical Physics* 24, no. 1 (January 1985): 34–36.

4. Časlav Brukner, "On the Quantum Measurement Problem," in *Quantum [Un]Speakables II: The Frontiers Collection*, ed. Reinhold Bertlmann and Anton Zeilinger (Cham, Switzerland: Springer International, 2017), 95–117.

5. Massimiliano Proietti et al., "Experimental Test of Local Observer Independence," *Science Advances* 5, no. 9 (20 September 2019): eaaw9832.

6. Kok-Wei Bong et al., "A Strong No-Go Theorem on the Wigner's Friend Paradox," *Nature Physics* 16 (17 August 2020): 1199–205; Musser, "Schrödinger's A.I. Could Test the Foundations of Reality."

7. D. Rauch et al., "Cosmic Bell Test Using Random Measurement Settings from High-Redshift Quasars," *Physical Review Letters* 121, no. 8 (24 August 2018): 080403.

8. Tim Maudlin, *Quantum Non-Locality and Relativity: Metaphysical Intimations of Modern Physics*, 2nd ed. (Malden, MA: Blackwell, 2002), 212–20.

9. Asher Peres, "Unperformed Experiments Have No Results," *American Journal of Physics* 46, no. 7 (July 1978): 745–47.

10. Christopher A. Fuchs, N. David Mermin, and Rüdiger Schack, "An Introduction to QBism with an Application to the Locality of Quantum Mechanics," *American Journal of Physics* 82, no. 8 (August 2014): 749–54.

11. Daniela Frauchiger and Renato Renner, "Quantum Theory Cannot Consistently Describe the Use of Itself," *Nature Communications* 9, no. 1 (18 September 2018): 823.

12. George Musser, "Watching the Watchmen: Demystifying the Frauchiger-Renner Experiment—Musings from Lidia del Rio and More at the 6th FQxI Meeting," *FQxI Blogs*, 24 December 2019, https://fqxi.org/community/forum/topic/3354.

13. Musser, "Watching the Watchmen."

14. Immanuel Kant, *Critique of Pure Reason*, ed. Vasilis Politis (London: J. M. Dent, 1993), 15.

15. Thomas S. Kuhn, *The Copernican Revolution: Planetary Astronomy in the Development of Western Thought* (Cambridge, MA: Harvard University Press, 1957), 150–53.

16. Don Howard, "Revisiting the Einstein-Bohr Dialogue," *iyyun: The Jerusalem Philosophical Quarterly* 56 (January 2007): 57–90.

17. David Bohm, *Quantum Theory* (New York: Dover, 1951), 26.

18. Max Jammer, *The Philosophy of Quantum Mechanics: The Interpretations of Quantum Mechanics in Historical Perspective* (New York: John Wiley and Sons, 1974), 200–202.

19. Barrett and Byrne, *Everett Interpretation of Quantum Mechanics*, 19–20.

20. Barrett and Byrne, *Everett Interpretation of Quantum Mechanics*, 17–18.

21. Everett, "Universal Wave Function," 64.

22. Everett, "Universal Wave Function," 98.

23. Everett, "Universal Wave Function," 68–77.

24. Barrett and Byrne, *Everett Interpretation of Quantum Mechanics*, 36–37.

25. Everett, "Universal Wave Function," 79–80.

26. Jeffrey A. Barrett, "Empirical Adequacy and the Availability of Reliable Records in Quantum Mechanics," *Philosophy of Science* 63, no. 1 (March 1996): 49–64.

27. Barrett and Byrne, *Everett Interpretation of Quantum Mechanics*, 50–54.

28. Barrett and Byrne, *Everett Interpretation of Quantum Mechanics*, 22–23.

29. David Wallace, "Everett and Structure," *Studies in History and Philosophy of Science Part B: Studies in History and Philosophy of Modern Physics* 34, no. 1 (March 2003): 87–105.

30. Richard A. Healey, "How Many Worlds?," *Noûs* 18, no. 4 (November 1984): 591; Barrett and Byrne, *Everett Interpretation of Quantum Mechanics*, 38.

31. David Albert and Barry Loewer, "Interpreting the Many Worlds Interpretation," *Synthese* 77, no. 2 (November 1988): 195–213.

32. Albert and Loewer, "Interpreting the Many Worlds Interpretation," 210–11.

33. Carlo Rovelli, "Relational Quantum Mechanics," *International Journal of Theoretical Physics* 35, no. 8 (August 1996): 1637–78.

34. Laura Candiotto, "The Reality of Relations," *Giornale di Metafisica* 39, no. 2 (2017): 537–51.

35. Sebastián Briceño and Stephen Mumford, "Relations All the Way Down? Against Ontic Structural Realism," in *The Metaphysics of Relations*, ed. Anna Marmodoro and David Yates (New York: Oxford University Press, 2016), 198–217.

36. Philip Goff, *Galileo's Error: Foundations for a New Science of Consciousness* (New York: Pantheon, 2019), 176–81.

37. Christopher A. Fuchs, Maximilian Schlosshauer, and Blake C. Stacey, "My Struggles with the Block Universe" (preprint, submitted 10 May 2014).

38. Richard Healey, "Quantum Theory: A Pragmatist Approach," *British Journal for the Philosophy of Science* 63, no. 4 (December 2012): 729–71.

39. Robert P. Crease and James Sares, "Interview with Physicist Christopher Fuchs," *Continental Philosophy Review* 54, no. 4 (December 2021): 541–61.

40. Guido Bacciagaluppi, "A Critic Looks at QBism," in *New Directions in the Philosophy of Science*, ed. Maria Carla Galavotti et al. (Cham, Switzerland: Springer International, 2014), 403–16; Jacques L. Pienaar, "A Quintet of Quandaries: Five No-Go Theorems for Relational Quantum Mechanics," *Foundations of Physics* 51 (4 October 2021): 97.

41. Ricardo Muciño, Elias Okon, and Daniel Sudarsky, "Assessing Relational Quantum Mechanics," *Synthese* 200, no. 5 (October 2022): 399.

42. Emily Adlam and Carlo Rovelli, "Information Is Physical: Cross-Perspective Links in Relational Quantum Mechanics" (preprint, submitted 24 March 2022).

6. MINDING THE UNIVERSE

1. Markus P. Müller, "Law Without Law: From Observer States to Physics via Algorithmic Information Theory," *Quantum* 4 (20 July 2020): 301.

2. Robert W. Smith, "Beyond the Galaxy: The Development of Extragalactic Astronomy 1885–1965, Part 1," *Journal for the History of Astronomy* 39, no. 1 (February 2008): 91–119.

3. Christopher J. Conselice et al., "The Evolution of Galaxy Number Density at $z < 8$ and Its Implications," *Astrophysical Journal* 830, no. 2 (2016): 83.

4. N. Aghanim et al., "Planck 2018 Results VI. Cosmological Parameters," *Astronomy & Astrophysics* 641 (2020): A6.

5. Mihran Vardanyan, Roberto Trotta, and Joseph Silk, "Applications of Bayesian Model Averaging to the Curvature and Size of the Universe," *Monthly Notices of the Royal Astronomical Society: Letters* 413, no. 1 (May 2011): L91–95.

6. Yashar Akrami et al., "The Search for the Topology of the Universe Has Just Begun" (preprint, submitted 20 October 2022).

7. Brian Greene, *The Hidden Reality: Parallel Universes and the Deep Laws of the Cosmos* (New York: Knopf, 2011), 44–45.

8. Greene, *Hidden Reality*, 54–56.

9. Greene, *Hidden Reality*, 27–35.

10. Fred C. Adams, "The Degree of Fine-Tuning in Our Universe—and Others," *Physics Reports* 807 (15 May 2019): 1–111.

11. Andrej B. Arbuzov, "Quantum Field Theory and the Electroweak Standard Model," in *Proceedings of the 2017 European School of High-Energy Physics*, ed. M. Mulders and G. Zanderighi (Geneva: CERN, 2018), sec. 3.11.

12. Nick Bostrom, *Anthropic Bias: Observation Selection Effects in Science and Philosophy* (New York: Routledge, 2002), 82–84.

13. John Locke, *An Essay Concerning Humane Understanding* (London: Awnsham and John Churchill, 1706), 226–27.

14. Locke, *Essay Concerning Humane Understanding*, 225.

15. Adam Elga, "Self-Locating Belief and the Sleeping Beauty Problem," *Analysis* 60, no. 2 (April 2000): 143–47.

16. Michele Piccione and Ariel Rubinstein, "On the Interpretation of Decision Problems with Imperfect Recall," *Games and Economic Behavior* 20, no. 1 (July 1997): 3–24.

17. M. Vittoria Levati, Matthias Uhl, and Ro'i Zultan, "Imperfect Recall and Time Inconsistencies: An Experimental Test of the Absentminded Driver 'Paradox,'" *International Journal of Game Theory* 43, no. 1 (23 April 2013): 65–88.

18. Charles T. Sebens and Sean M. Carroll, "Self-Locating Uncertainty and the Origin of Probability in Everettian Quantum Mechanics," *British Journal for the Philosophy of Science* 69, no. 1 (March 2018): 25–74.

19. Anthony Aguirre and Max Tegmark, "Born in an Infinite Universe: A Cosmological Interpretation of Quantum Mechanics," *Physical Review D* 84, no. 10 (3 November 2011): 105002.

20. René Descartes, *Meditations on First Philosophy*, trans. Michael Moriarty (New York: Oxford University Press, 2008), 16.

21. Locke, *Essay Concerning Humane Understanding*, 228.

22. Nick Bostrom, "Are We Living in a Computer Simulation?," *Philosophical Quarterly* 53, no. 211 (April 2003): 243–55.

23. Piero Madau and Mark Dickinson, "Cosmic Star-Formation History," *Annual Review of Astronomy and Astrophysics* 52, no. 1 (August 2014): 415–86.

24. Sean M. Carroll, "The Cosmic Origins of Time's Arrow," *Scientific American* 298, no. 6 (June 2008): 48–53, 56.
25. Andreas Albrecht and Lorenzo Sorbo, "Can the Universe Afford Inflation?," *Physical Review D* 70, no. 6 (21 September 2004): 063528.
26. Sean M. Carroll, "Why Boltzmann Brains Are Bad" (preprint, submitted 2 February 2017).
27. Paul J. Steinhardt, "The Inflation Debate," *Scientific American* 304, no. 4 (April 2011): 36–43.
28. Raphael Bousso and Ben Freivogel, "A Paradox in the Global Description of the Multiverse," *Journal of High Energy Physics* 2007, no. 6 (June 2007): 018.
29. Cian Dorr and Frank Arntzenius, "Self-Locating Priors and Cosmological Measures," in *The Philosophy of Cosmology* (Cambridge: Cambridge University Press, 2017), 396–428.
30. Scott Aaronson, "The Ghost in the Quantum Turing Machine," in *The Once and Future Turing*, ed. S. Barry Cooper and Andrew Hodges (New York: Cambridge University Press, 2016), sec. 7.
31. R. J. Solomonoff, "A Formal Theory of Inductive Inference. Part I," *Information and Control* 7, no. 1 (March 1964): 1–22.
32. David J. Chalmers, "The Virtual and the Real," *Disputatio* 9, no. 46 (2017): 309–52.

7. LEVELS OF REALITY

1. Bertrand Russell, "On the Notion of Cause," *Proceedings of the Aristotelian Society* 13 (1912): 1–26.
2. William Seager, *Natural Fabrications: Science, Emergence and Consciousness* (New York: Springer, 2012), 183.
3. Kirsty L. Spalding et al., "Retrospective Birth Dating of Cells in Humans," *Cell* 122, no. 1 (15 July 2005): 133–43.
4. Silvan S. Schweber, "Physics, Community and the Crisis in Physical Theory," *Physics Today* 46, no. 11 (November 1993): 34–40.
5. Albert Einstein, *The Meaning of Relativity* (Princeton, NJ: Princeton University Press, 2005), 55–56.
6. Samuel Alexander, *Space, Time, and Deity: The Gifford Lectures at Glasgow, 1916–1918* (Gloucester, MA: Peter Smith, 1979), 8.
7. Graham Oddie, "Armstrong on the Eleatic Principle and Abstract Entities," *Philosophical Studies* 41, no. 2 (March 1982): 285–95.
8. Eugene Paul Wigner, "Remarks on the Mind-Body Question," in *The Scientist Speculates: An Anthology of Partly-Baked Ideas*, ed. Irving John Good (London: Heinemann, 1962), 294.
9. Alexander, *Space, Time, and Deity*, 45–47.
10. Alexander, *Space, Time, and Deity*, 14n2.
11. Philip W. Anderson, "More Is Different," *Science* 177, no. 4 (4 August 1972): 393–96.

12. Núria Muñoz Garganté, "A Physicist's Road to Emergence: Revisiting the Story of 'More Is Different'" (lecture, Max Planck Institute for the History of Science, Berlin, 3 September 2019).

13. Philip W. Anderson, "More Is Different—One More Time," in *More Is Different*, ed. N. Phuan Ong and Ravin N. Bhatt (Princeton, NJ: Princeton University Press, 2001), 1–8.

14. Judea Pearl, *Causality*, 2nd ed. (New York: Cambridge University Press, 2009), 419–20.

15. Jenann T. Ismael, *How Physics Makes Us Free* (New York: Oxford University Press, 2016), chap. 5.

16. Huw Price, "Causal Perspectivalism," in *Causation, Physics, and the Constitution of Reality*, ed. Huw Price and Richard Corry (New York: Oxford University Press, 2007), 250–92.

17. Daniel M. Hausman, *Causal Asymmetries* (New York: Cambridge University Press, 1998), chap. 5.

18. Sewall Wright, "Correlation and Causation," *Journal of Agricultural Research* 20, no. 7 (3 January 1921): 557–85.

19. Pearl, *Causality*, 423.

20. Pearl, *Causality*, 83–85.

21. Judea Pearl and Dana Mackenzie, *The Book of Why: The New Science of Cause and Effect* (New York: Basic Books, 2018), 293–96.

22. Erik P. Hoel, "Causal Structure Across Scales" (lecture, Araya Brain Imaging, Tokyo, 27 April 2018).

23. Veeresh Taranalli, Hironori Uchikawa, and Paul H. Siegel, "Channel Models for Multi-Level Cell Flash Memories Based on Empirical Error Analysis," *IEEE Transactions on Communications* 64, no. 8 (August 2016): 3169–81.

24. John Archibald Wheeler, "Pregeometry: Motivations and Prospects," in *Quantum Theory and Gravitation*, ed. A. R. Marlow (New York: Academic Press, 1980), 6.

25. Irina Higgins et al., "SCAN: Learning Hierarchical Compositional Visual Concepts" (preprint, submitted 11 July 2017).

26. Roderick M. Chisholm, *Human Freedom and the Self* (Lawrence: Dept. of Philosophy, University of Kansas, 1964).

27. Luca Bombelli et al., "Space-Time as a Causal Set," *Physical Review Letters* 59, no. 5 (3 August 1987): 521–24.

28. Daniel C. Dennett, *Consciousness Explained* (Boston: Little, Brown, 1991), 107.

29. Giulio Tononi et al., "Only What Exists Can Cause: An Intrinsic View of Free Will" (preprint, submitted 4 June 2022).

30. Ian Durham, "A Formal Model for Adaptive Free Choice in Complex Systems," *Entropy* 22, no. 5 (19 May 2020): 568.

31. Andrea Lavazza and Silvia Inglese, "Operationalizing and Measuring (a Kind of) Free Will (and Responsibility). Towards a New Framework for Psychology, Ethics, and Law," *Rivista Internazionale di Filosofia e Psicologia* 6, no. 1 (17 April 2015): 37–55.

32. Jonathan Barrett and Nicolas Gisin, "How Much Measurement Independence Is Needed to Demonstrate Nonlocality?," *Physical Review Letters* 106, no. 10 (10 March 2011): 100406.

33. Lars Marstaller, Arend Hintze, and Christoph Adami, "The Evolution of Representation in Simple Cognitive Networks," *Neural Computation* 25, no. 8 (August 2013): 2079–107.

34. Sarah Scoles, "NASA's Space Crash Succeeded in Forcing Asteroid onto New Path," *New York Times*, 12 October 2022.

35. Ismael, *How Physics Makes Us Free*, 105–106.

8. TIME AND SPACE

1. Ruth S. Ogden, "The Passage of Time During the UK Covid-19 Lockdown," *PLoS ONE* 15, no. 7 (6 July 2020): e02358871.

2. Stefanie Hüttermann, Benjamin Noël, and Daniel Memmert, "Evaluating Erroneous Offside Calls in Soccer," *PLoS ONE* 12, no. 3 (23 March 2017): e0174358.

3. Dean Buonomano, "Time, the Brain, and Consciousness" (lecture, The Science of Consciousness 2020, 15 September 2020).

4. Ralf Steinmetz, "Human Perception of Jitter and Media Synchronization," *IEEE Journal on Selected Areas in Communications* 14, no. 1 (January 1996): 61–72.

5. Barbara Tversky, *Mind in Motion* (New York: Basic Books, 2019), 80–82.

6. Mary B. Hesse, *Forces and Fields: The Concept of Action at a Distance in the History of Physics* (Mineola, NY: Dover, 2005), 39–40.

7. Plato, "Parmenides," in *Readings in Ancient Greek Philosophy: From Thales to Aristotle*, ed. S. Marc Cohen, Patricia Curd, and C. D. C. Reeve (Cambridge, MA: Hackett, 1995), 433.

8. Carlo Rovelli, *Quantum Gravity* (New York: Cambridge University Press, 2004), 84–87.

9. Richard A. Healey, "Can Physics Coherently Deny the Reality of Time?," in *Time, Reality and Experience*, ed. Craig Callender (New York: Cambridge University Press, 2002), 293–316.

10. Lee Smolin, *Time Reborn: From the Crisis in Physics to the Future of the Universe* (New York: Houghton Mifflin Harcourt, 2013).

11. Nathan Seiberg, "Emergent Spacetime," in *The Quantum Structure of Space and Time*, ed. David J. Gross, Marc Henneaux, and Alexander Sevrin (Hackensack, NJ: World Scientific, 2007), 162–213.

12. Smolin, *Time Reborn*, chap. 15.

13. John Archibald Wheeler, Papers 52:140, Relativity Notebook #14, 27 January 1967, American Philosophical Library, Philadelphia.

14. Christopher J. Isham, "Canonical Quantum Gravity and the Problem of Time," in *Integrable Systems, Quantum Groups, and Quantum Field Theories*, ed. L. A. Ibort and M. A. Rodríguez (Dordrecht: Springer Netherlands, 1993), 157–287.

15. Donald Salisbury, "Toward a Quantum Theory of Gravity: Syracuse 1949–1962," in *The Renaissance of General Relativity in Context*, ed. Alexander S. Blum, Roberto Lalli, and Jürgen Renn (Cham, Switzerland: Springer International, 2020), 221–55.

16. John Stachel, "The Other Einstein: Einstein Contra Field Theory," *Science in Context* 6, no. 1 (Spring 1993): 285.

17. Bryce S. DeWitt, "Quantum Theory of Gravity. I. The Canonical Theory," *Physical Review* 160, no. 5 (25 August 1967): 221–55.

18. Don N. Page and William K. Wootters, "Evolution Without Evolution: Dynamics Described by Stationary Observables," *Physical Review D* 27, no. 12 (15 June 1983): 2885–92.

19. Ekaterina Moreva et al., "Time from Quantum Entanglement: An Experimental Illustration," *Physical Review A* 89, no. 5 (20 May 2014): 052122.

20. Carlo Rovelli, "Forget Time" (preprint, submitted 23 March 2009).

21. Carlo Rovelli, *What Is Time? What Is Space?*, trans. J. C. van den Berg (Rome: Di Renzo Editore, 2006), 52.

22. Rovelli, *Quantum Gravity*, 140–44.

23. Zafeirios Fountas et al., "A Predictive Processing Model of Episodic Memory and Time Perception," *Neural Computation* 34, no. 7 (16 June 2022).

24. Ludwig Boltzmann, *Lectures on Gas Theory*, trans. Stephen G. Brush (Berkeley: University of California Press, 1964), 447.

25. Arseni Goussev et al., "Loschmidt Echo," *Scholarpedia* 7, no. 8 (27 June 2012): 11687.

26. Sean M. Carroll, "Cosmic Origins of Time's Arrow," *Scientific American* 298, no. 6 (June 2008): 48–53, 56.

27. Carlo Rovelli, "Is Time's Arrow Perspectival?," in *The Philosophy of Cosmology*, ed. Khalil Chamcham et al. (New York: Cambridge University Press, 2017), 285–96.

28. Leonard Mlodinow and Todd A. Brun, "Relation Between the Psychological and Thermodynamic Arrows of Time," *Physical Review E* 89, no. 5 (May 2014): 052102.

29. Meir Hemmo and Orly Shenker, "Can the Past Hypothesis Explain the Psychological Arrow of Time?," in *Statistical Mechanics and Scientific Explanation*, ed. Valia Allori (Singapore: World Scientific, 2020), 255–87.

30. Huw Price, *Time's Arrow and Archimedes' Point: New Directions for the Physics of Time* (New York: Oxford University Press, 1996), 106–109.

31. Lorenzo Maccone, "Quantum Solution to the Arrow-of-Time Dilemma," *Physical Review Letters* 103, no. 8 (21 August 2009): 080401.

32. Ken B. Wharton, "Time-Symmetric Boundary Conditions and Quantum Foundations," *Symmetry* 2, no. 1 (March 2010): 272–83.

33. Barry Dainton, "Temporal Consciousness," *Stanford Encyclopedia of Philosophy*, last modified 28 June 2017, https://plato.stanford.edu/entries/consciousness-temporal/.

34. John Locke, *An Essay Concerning Humane Understanding* (London: Awnsham and John Churchill, 1706), 111–12.

35. William James, *The Principles of Psychology* (New York: Holt, 1890), 609.

36. Edmund R. Clay, *The Alternative: A Study in Psychology* (London: Macmillan, 1882), 168.
37. Holly K. Andersen and Rick Grush, "A Brief History of Time-Consciousness: Historical Precursors to James and Husserl," *Journal of the History of Philosophy* 47, no. 2 (April 2009): 277–307.
38. Stanislas Dehaene, *Consciousness and the Brain: Deciphering How the Brain Codes Our Thoughts* (New York: Penguin, 2014), 100–104.
39. Dehaene, *Consciousness and the Brain*, 63–64.
40. Ryota Kanai, "We Need Conscious Robots," *Nautilus* 47 (27 April 2017), https://nautil.us/we-need-conscious-robots-236579/.
41. Maxwell Nye et al., "Show Your Work: Scratchpads for Intermediate Computation with Language Models" (preprint, submitted 30 November 2021).
42. David J. Chalmers, "Could a Large Language Model Be Conscious?" (lecture, Conference on Neural Information Processing Systems, New Orleans, 28 November 2022).
43. Rick Grush, "Brain Time and Phenomenological Time," in *Cognition and the Brain*, ed. Andrew Brook and Kathleen Akins (New York: Cambridge University Press, 2005), 160–207.
44. Kristie Miller, Alex Holcombe, and Andrew James Latham, "Temporal Phenomenology: Phenomenological Illusion Versus Cognitive Error," *Synthese* 197, no. 2 (February 2020): 751–71.
45. Murray Gell-Mann and James B. Hartle, "Quantum Mechanics in the Light of Quantum Cosmology," in *Proceedings of the 3rd International Symposium Foundations of Quantum Mechanics in the Light of New Technology*, ed. Shun'ichi Kobayashi et al. (Tokyo: Physical Society of Japan, 1990), 321–43.
46. James B. Hartle, "The Physics of Now," *American Journal of Physics* 73, no. 2 (February 2005): 101–109.
47. Jakob Hohwy, Bryan Paton, and Colin Palmer, "Distrusting the Present," *Phenomenology and the Cognitive Sciences* 15, no. 3 (September 2016): 315–35.
48. Craig Callender, *What Makes Time Special* (New York: Oxford University Press, 2017), 247–55.
49. Daniele Oriti, "Levels of Spacetime Emergence in Quantum Gravity," in *Philosophy Beyond Spacetime*, ed. Christian Wüthrich, Baptiste Le Bihan, and Nick Huggett (New York: Oxford University Press, 2021), 16–40.
50. Mark Van Raamsdonk, "Building Up Spacetime with Quantum Entanglement," *General Relativity and Gravitation* 42, no. 10 (19 June 2010): 2323–29; Michael Heller and Wieslaw Sasin, "Towards Noncommutative Quantization of Gravity" (preprint, submitted 1 December 1997).
51. Garry Kasparov, *Deep Thinking: Where Machine Intelligence Ends and Human Creativity Begins* (New York: PublicAffairs, 2017), 80, 137, 190–91.
52. Edvard I. Moser, Emilio Kropff, and May-Britt Moser, "Place Cells, Grid Cells, and the Brain's Spatial Representation System," *Annual Review of Neuroscience* 31 (2008): 69–89.
53. Dehaene, *Consciousness and the Brain*, 61–62.

54. William James, "The Spatial Quale," *Journal of Speculative Philosophy* 13, no. 1 (January 1879): 75.

55. Oliver Sacks, *The Man Who Mistook His Wife for A Hat: And Other Clinical Tales* (New York: Simon and Schuster, 1998), 77–79.

56. Scott Aaronson, "Why I Am Not an Integrated Information Theorist (or, The Unconscious Expander)," *Shtetl-Optimized* (blog), 24 May 2014, https://www.scottaaronson.com/blog/?p=1799.

57. Andrew M. Haun and Giulio Tononi, "Why Does Space Feel the Way It Does? Towards a Principled Account of Spatial Experience," *Entropy* 21, no. 12 (December 2019): 1160.

58. Andrew M. Haun, "A Causal Account of Spatial Experience: IIT and the Visual Field" (lecture, Mathematical Consciousness Science, 13 August 2020).

59. Pankaj Mehta and David J. Schwab, "An Exact Mapping Between the Variational Renormalization Group and Deep Learning" (preprint, submitted 14 October 2014).

60. Brian Greene, *The Hidden Reality: Parallel Universes and the Deep Laws of the Cosmos* (New York: Knopf, 2011), 262–69.

61. Wen-Cong Gan and Fu-Wen Shu, "Holography as Deep Learning," *International Journal of Modern Physics D* 26, no. 12 (October 2017): 1743020.

62. Koji Hashimoto et al., "Deep Learning and the AdS/CFT Correspondence," *Physical Review D* 98, no. 4 (27 August 2018): 046019.

63. Vitaly Vanchurin, "The World as a Neural Network," *Entropy* 22, no. 11 (26 October 2020): 1210.

64. Stephon Alexander et al., "The Autodidactic Universe" (preprint, submitted 29 March 2021).

65. Richard A. Watson and Eörs Szathmáry, "How Can Evolution Learn?," *Trends in Ecology and Evolution* 31, no. 2 (February 2016): 147–57.

66. Colin McGinn, "Consciousness and Space," *Journal of Consciousness Studies* 2, no. 3 (1 March 1995): 220–30.

67. David Bohm, "Time, the Implicate Order and Pre-Space," in *Physics and the Ultimate Significance of Time*, ed. David Ray Griffin (Albany: State University of New York Press, 1986), 177–208.

68. Donald D. Hoffman, *The Case Against Reality: Why Evolution Hid the Truth from Our Eyes* (New York: W. W. Norton, 2019).

69. Immanuel Kant, *Critique of Pure Reason*, ed. Vasilis Politis (London: J. M. Dent, 1993), 49–54.

70. Donald D. Hoffman, "The Origin of Time in Conscious Agents," *Cosmology* 18 (November 2014): 494–520.

EPILOGUE: IS IT REALLY SO HARD?

1. Owen Flanagan, *The Science of the Mind* (Cambridge, MA: MIT Press, 1991), 313–14.

2. Steven Pinker, *How the Mind Works* (New York: Penguin Random House, 2015), 561–65; Noam Chomsky, *The Essential Chomsky* (New York: New Press, 2008), 235–44.

3. Carlo Rovelli, "Consciousness, Time, Quantum" (lecture, The Science of Consciousness, 15 September 2020); Carlo Rovelli, *Helgoland* (New York: Penguin, 2021), 184–86.

4. Emily Adlam and Carlo Rovelli, "Information Is Physical: Cross-Perspective Links in Relational Quantum Mechanics" (preprint, submitted 24 March 2022).

5. M. D. Beni, "A Structuralist Defence of the Integrated Information Theory of Consciousness," *Journal of Consciousness Studies* 25, no. 9–10 (2018): 75–98.

6. David Balduzzi and Giulio Tononi, "Qualia: The Geometry of Integrated Information" *PLoS Computational Biology* 5, no. 8 (August 2009): e1000462.

7. Daniel C. Dennett, "Expecting Ourselves to Expect: The Bayesian Brain as a Projector," *Behavioral and Brain Sciences* 36, no. 3 (June 2013): 209–10.

8. Andy Clark, Karl J. Friston, and Sam Wilkinson, "Bayesing Qualia: Consciousness as Inference, Not Raw Datum," *Journal of Consciousness Studies* 26 (2019): 19–33.

9. Kristjan Loorits, "Structural Qualia: A Solution to the Hard Problem of Consciousness," *Frontiers in Psychology* 5 (18 March 2014): 237.

10. Daniel C. Dennett, *Consciousness Explained* (Boston: Little, Brown, 1991), 134–38.

11. Keith Frankish, "Illusionism as a Theory of Consciousness," *Journal of Consciousness Studies* 23, no. 11–12 (2016): 11–39.

12. Kristjan Loorits, "Qualities in the World, in Science, and in Consciousness," *Journal of Consciousness Studies* 29, no. 11 (2022): 108–30.

13. Richard P. Stanley, "Qualia Space," *Journal of Consciousness Studies* 6, no. 1 (1 January 1999): 49–60.

14. Naotsugu Tsuchiya, Shigeru Taguchi, and Hayato Saigo, "Using Category Theory to Assess the Relationship Between Consciousness and Integrated Information Theory," *Neuroscience Research* 107 (June 2016): 1–7.

15. Frank Jackson, "Epiphenomenal Qualia," *Philosophical Quarterly* 32, no. 127 (April 1982): 127.

16. John Locke, *An Essay Concerning Humane Understanding* (London: Awnsham and John Churchill, 1706), 68, 365.

17. Isaac Newton, *Opticks: or, A Treatise of the Reflections, Refractions, Inflections and Colours of Light* (London: Royal Society, 1718), 134–37.

18. Armin Saysani, Michael C. Corballis, and Paul M. Corballis, "Colour Envisioned: Concepts of Colour in the Blind and Sighted," *Visual Cognition* 26, no. 5 (May 2018): 382–92.

19. Helen Keller, *The World I Live In* (New York: Century, 1908), 106.

20. Mostafa Abdou et al., "Can Language Models Encode Perceptual Structure Without Grounding? A Case Study in Color" (preprint, submitted 13 September 2021).

21. Ron Chrisley, "Interactive Empiricism: The Philosopher in the Machine," in *Philosophy of Engineering* (London: Royal Academy of Engineering, 2010), 66–71.

22. Yuichi Yamashita and Jun Tani, "Spontaneous Prediction Error Generation in Schizophrenia," *PLoS ONE* 7, no. 5 (30 May 2012): e37843; Hayato Idei et al., "A Neurorobotics Simulation of Autistic Behavior Induced by Unusual Sensory Precision," *Computational Psychiatry* 2 (2 December 2018): 164–82.

23. Giuseppe Carleo et al., "Machine Learning and the Physical Sciences," *Reviews of Modern Physics* 91, no. 4 (5 December 2019): 045002.

24. M. Schmidt and H. Lipson, "Distilling Free-Form Natural Laws from Experimental Data," *Science* 324, no. 5923 (3 April 2009): 81–85.

25. Raban Iten et al., "Discovering Physical Concepts with Neural Networks," *Physical Review Letters* 124, no. 1 (10 January 2020): 010508; Mark Stalzer and Chao Ju, "Automated Rediscovery of the Maxwell Equations," *Applied Sciences* 9, no. 14 (19 July 2019): 2899; Samuel H. Rudy et al., "Data-Driven Discovery of Partial Differential Equations," *Science Advances* 3, no. 4 (April 2017): e1602614; Silviu-Marian Udrescu and Max Tegmark, "AI Feynman: A Physics-Inspired Method for Symbolic Regression," *Science Advances* 6, no. 16 (15 April 2020): eaay2631; Alireza Seif, Mohammad Hafezi, and Christopher Jarzynski, "Machine Learning the Thermodynamic Arrow of Time," *Nature Physics* 17, no. 1 (January 2021): 105–13.

26. Vishnu Jejjala, Arjun Kar, and Onkar Parrikar, "Deep Learning the Hyperbolic Volume of a Knot," *Physics Letters B* 799 (10 December 2019): 135033.

27. Toby S. Cubitt, Jens Eisert, and Michael M. Wolf, "Extracting Dynamical Equations from Experimental Data Is NP Hard," *Physical Review Letters* 108, no. 12 (23 March 2012): 120503.

28. Gary Marcus and Ernest Davis, "Are Neural Networks About to Reinvent Physics?," *Nautilus* 78 (21 November 2019), https://nautil.us/are-neural-networks-about -to-reinvent-physics-237619/.

29. Steven Johnson, *Where Good Ideas Come From: The Natural History of Innovation* (New York: Penguin Group, 2010), chap. 5.

30. Mary B. Hesse, *Forces and Fields: The Concept of Action at a Distance in the History of Physics* (Mineola, NY: Dover, 2005), 126–31.

31. Hendrik Poulsen Nautrup et al., "Operationally Meaningful Representations of Physical Systems in Neural Networks," *Machine Learning: Science and Technology* 3, no. 4 (December 2022): 045025.

32. Bart Selman, "Dirty Secrets of Artificial Intelligence" (lecture, Foundational Questions Institute Fifth International Conference, Banff, Canada, 18 August 2016).

33. Terence Tao, "The Erdős Discrepancy Problem," *Discrete Analysis* 1 (28 February 2016).

34. Daniel C. Dennett, *Darwin's Dangerous Idea* (New York: Simon and Schuster, 1996), 381–83.

35. Margaret A. Boden, *Mind as Machine: A History of Cognitive Science*, 2 vols. (New York: Oxford University Press, 2006), 173–75.

36. Chomsky, *Essential Chomsky*, 34–38; Pinker, *How the Mind Works*, 124–25.

37. Zohar Z. Bronfman, Simona Ginsburg, and Eva Jablonka, "The Transition to Minimal Consciousness Through the Evolution of Associative Learning," *Frontiers in Psychology* 7 (2016): 1954.

BIBLIOGRAPHY

Aaronson, Scott. "The Ghost in the Quantum Turing Machine." In *The Once and Future Turing*, edited by S. Barry Cooper and Andrew Hodges, 193–296. New York: Cambridge University Press, 2016.

———. "Why I Am Not an Integrated Information Theorist (or, The Unconscious Expander)." *Shtetl-Optimized* (blog), 24 May 2014. https://www.scottaaronson.com/blog/?p=1799.

Abbott, B. P., R. Abbott, T. D. Abbott, M. R. Abernathy, F. Acernese, K. Ackley, C. Adams et al. "Observation of Gravitational Waves from a Binary Black Hole Merger." *Physical Review Letters* 116, no. 6 (12 February 2016): 061102.

Abdou, Mostafa, Artur Kulmizev, Daniel Hershcovich, Stella Frank, Ellie Pavlick, and Anders Søgaard. "Can Language Models Encode Perceptual Structure Without Grounding? A Case Study in Color." Preprint, submitted 13 September 2021. https://arxiv.org/abs/2109.06129.

Ackley, David H., Geoffrey E. Hinton, and Terrence J. Sejnowski. "A Learning Algorithm for Boltzmann Machines." *Cognitive Science* 9, no. 1 (January 1985): 147–69.

Adams, Fred C. "The Degree of Fine-Tuning in Our Universe—and Others." *Physics Reports* 807 (15 May 2019): 1–111.

Adlam, Emily, and Carlo Rovelli. "Information Is Physical: Cross-Perspective Links in Relational Quantum Mechanics." Preprint, submitted 24 March 2022. https://arxiv.org/abs/2203.13342.

Aghanim, N., Y. Akrami, M. Ashdown, J. Aumont, C. Baccigalupi, M. Ballardini, A. J. Banday et al. "Planck 2018 Results VI. Cosmological Parameters." *Astronomy & Astrophysics* 641 (2020): A6.

Aguirre, Anthony, and Max Tegmark. "Born in an Infinite Universe: A Cosmological Interpretation of Quantum Mechanics." *Physical Review D* 84, no. 10 (3 November 2011): 105002.

Akrami, Yashar, Craig J. Copi, Johannes R. Eskilt, Andrew H. Jaffe, Arthur Kosowsky, Pip Petersen, Glenn D. Starkman et al. "The Search for the Topology of the Universe Has Just Begun." Preprint, submitted 20 October 2022. https://arxiv.org/abs/2210.11426.

Albantakis, Larissa, Arend Hintze, Christof Koch, Christoph Adami, and Giulio Tononi. "Evolution of Integrated Causal Structures in Animats Exposed to Environments of Increasing Complexity." *PLoS Computational Biology* 10, no. 12 (18 December 2014): e1003966.

Albert, David, and Barry Loewer. "Interpreting the Many Worlds Interpretation." *Synthese* 77, no. 2 (November 1988): 195–213.

Albrecht, Andreas, and Lorenzo Sorbo. "Can the Universe Afford Inflation?" *Physical Review D* 70, no. 6 (21 September 2004): 063528.

Alexander, Samuel. *Space, Time, and Deity: The Gifford Lectures at Glasgow, 1916–1918.* Gloucester, MA: Peter Smith, 1979.

Alexander, Stephon, William J. Cunningham, Jaron Lanier, Lee Smolin, Stefan Stanojevic, Michael W. Toomey, and Dave Wecker. "The Autodidactic Universe." Preprint, submitted 29 March 2021. https://arxiv.org/abs/2104.03902.

Amari, Shun-Ichi. "Neural Theory of Association and Concept-Formation." *Biological Cybernetics* 26, no. 3 (17 May 1977): 175–85.

Amit, Daniel J., Hanoch Gutfreund, and Haim Sompolinsky. "Spin-Glass Models of Neural Networks." *Physical Review A* 32, no. 2 (August 1985): 1007–18.

Andersen, Holly K., and Rick Grush. "A Brief History of Time-Consciousness: Historical Precursors to James and Husserl." *Journal of the History of Philosophy* 47, no. 2 (April 2009): 277–307.

Anderson, James A., and Edward Rosenfeld, eds. "Geoffrey E. Hinton." In *Talking Nets: An Oral History of Neural Networks,* 361–84. Cambridge, MA: MIT Press, 2000.

Anderson, Philip W. "More Is Different." *Science* 177, no. 4 (4 August 1972): 393–96.

———. "More Is Different—One More Time." In *More Is Different,* edited by N. Phuan Ong and Ravin N. Bhatt, 1–8. Princeton, NJ: Princeton University Press, 2001.

Arbuzov, Andrej B. "Quantum Field Theory and the Electroweak Standard Model." In *Proceedings of the 2017 European School of High-Energy Physics,* edited by M. Mulders and G. Zanderighi, 1–35. Geneva: CERN, 2018.

Arena, A., R. Comolatti, S. Thon, A. G. Casali, and J. F. Storm. "General Anesthesia Disrupts Complex Cortical Dynamics in Response to Intracranial Electrical Stimulation in Rats." *eNeuro* 8, no. 4 (5 August 2021): ENEURO.0343-20.2021.

Bacciagaluppi, Guido. "A Critic Looks at QBism." In *New Directions in the Philosophy of Science,* edited by Maria Carla Galavotti, Dennis Dieks, Wenceslao J. Gonzalez, Stephan Hartmann, Thomas Uebel, and Marcel Weber, 403–16. Cham, Switzerland: Springer International, 2014.

Bain, Alexander. *Mind and Body: The Theories of Their Relation.* New York: D. Appleton, 1875.

Balduzzi, David, and Giulio Tononi. "Qualia: The Geometry of Integrated Information." *PLoS Computational Biology* 5, no. 8 (August 2009): e1000462.

Ballard, Dana H., Geoffrey E. Hinton, and Terrence J. Sejnowski. "Parallel Visual Computation." *Nature* 306, no. 5938 (3 November 1983): 21–26.

Ballentine, Leslie E. "A Meeting with Wigner." *Foundations of Physics* 49, no. 8 (August 2019): 783–85.

Barrett, Jeffrey A. "Empirical Adequacy and the Availability of Reliable Records in Quantum Mechanics." *Philosophy of Science* 63, no. 1 (March 1996): 49–64.

Barrett, Jeffrey A., and Peter Byrne, eds. *The Everett Interpretation of Quantum Mechanics*. Princeton, NJ: Princeton University Press, 2012.

Barrett, Jonathan, and Nicolas Gisin. "How Much Measurement Independence Is Needed to Demonstrate Nonlocality?" *Physical Review Letters* 106, no. 10 (10 March 2011): 100406.

Bayne, Tim. "On the Axiomatic Foundations of the Integrated Information Theory of Consciousness." *Neuroscience of Consciousness* 2018, no. 1 (29 June 2018): 159.

Bayne, Tim, Jakob Hohwy, and Adrian M. Owen. "Are There Levels of Consciousness?" *Trends in Cognitive Sciences* 20, no. 6 (June 2016): 405–13.

Bayne, Tim, Anik K. Seth, and Marcello Massimini. "Are There Islands of Awareness?" *Trends in Neurosciences* 43, no. 1 (January 2020): 6–16.

Behrman, Elizabeth C., K. Gaddam, James E. Steck, and S. R. Skinner. "Microtubules as a Quantum Hopfield Network." In *The Emerging Physics of Consciousness*, edited by Jack Tuszyński, 351–70. Heidelberg: Springer Berlin Heidelberg, 2006.

Behrman, Elizabeth C., J. Niemel, James E. Steck, and S. R. Skinner. "A Quantum Dot Neural Network." In *Proceedings of the Fourth Workshop on Physics of Computation*, edited by Tommaso Toffoli, Michael Biafore, and João Leão, 22–24. Cambridge, MA: New England Complex Systems Institute, 1996.

Behrman, Elizabeth C., and James E. Steck. "Multiqubit Entanglement of a General Input State." *Quantum Information and Computation* 13, no. 1/2 (2013): 36–53.

Belkin, Mikhail, Daniel Hsu, Siyuan Ma, and Soumik Mandal. "Reconciling Modern Machine-Learning Practice and the Classical Bias-Variance Trade-Off." *Proceedings of the National Academy of Sciences of the United States of America* 116, no. 32 (6 August 2019): 15849–54.

Bell, John S. *Speakable and Unspeakable in Quantum Mechanics*. New York: Cambridge University Press, 2004.

Beni, M. D. "A Structuralist Defence of the Integrated Information Theory of Consciousness." *Journal of Consciousness Studies* 25, no. 9–10 (2018): 75–98.

Beniaguev, David, Idan Segev, and Michael London. "Single Cortical Neurons as Deep Artificial Neural Networks." *Neuron* 109, no. 17 (1 September 2021): 2727–39.e3.

Biamonte, Jacob, Peter Wittek, Nicola Pancotti, Patrick Rebentrost, Nathan Wiebe, and Seth Lloyd. "Quantum Machine Learning." *Nature* 549, no. 7671 (13 September 2017): 195–202.

Boden, Margaret A. *Mind as Machine: A History of Cognitive Science*. 2 vols. New York: Oxford University Press, 2006.

Bohm, David. *Quantum Theory*. New York: Dover, 1951.

———. "Time, the Implicate Order and Pre-Space." In *Physics and the Ultimate Significance of Time*, edited by David Ray Griffin, 177–208. Albany: State University of New York Press, 1986.

Boltzmann, Ludwig. *Lectures on Gas Theory*. Translated by Stephen G. Brush. Berkeley: University of California Press, 1964.

Boly, Mélanie. "Are the Neural Correlates of Consciousness (Mostly) in the Front or in the Back of the Cerebral Cortex?" Lecture, Association of the Scientific Study of Consciousness, Kraków, Poland, 18 June 2018.

Bombelli, Luca, Joohan Lee, David A. Meyer, and Rafael D. Sorkin. "Space-Time as a Causal Set." *Physical Review Letters* 59, no. 5 (3 August 1987): 521–24.

Bong, Kok-Wei, Aníbal Utreras-Alarcón, Farzad Ghafari, Yeong-Cherng Liang, Nora Tischler, Eric G. Cavalcanti, Geoff J. Pryde, and Howard M. Wiseman. "A Strong No-Go Theorem on the Wigner's Friend Paradox." *Nature Physics* 16 (17 August 2020): 1199–205.

Bostrom, Nick. *Anthropic Bias: Observation Selection Effects in Science and Philosophy.* New York: Routledge, 2002.

———. "Are We Living in a Computer Simulation?" *Philosophical Quarterly* 53, no. 211 (April 2003): 243–55.

Bousso, Raphael, and Ben Freivogel. "A Paradox in the Global Description of the Multiverse." *Journal of High Energy Physics* 2007, no. 6 (June 2007): 018.

Briceño, Sebastián, and Stephen Mumford. "Relations All the Way Down? Against Ontic Structural Realism." In *The Metaphysics of Relations*, edited by Anna Marmodoro and David Yates, 198–217. New York: Oxford University Press, 2016.

Bricken, Trenton, and Cengiz Pehlevan. "Attention Approximates Sparse Distributed Memory." Preprint, submitted 10 November 2021. https://arxiv.org/abs/2111.05498.

Bronfman, Zohar Z., Simona Ginsburg, and Eva Jablonka. "The Transition to Minimal Consciousness Through the Evolution of Associative Learning." *Frontiers in Psychology* 7 (2016): 1954.

Brukner, Časlav. "On the Quantum Measurement Problem." In *Quantum [Un]Speakables II: The Frontiers Collection*, edited by Reinhold Bertlmann and Anton Zeilinger, 95–117. Cham, Switzerland: Springer International, 2017.

Buonomano, Dean. "Time, the Brain, and Consciousness." Lecture, The Science of Consciousness 2020, 15 September 2020.

Cable, Hugo, and Kavan Modi. "Harness Quantum Noise to Unlock Quantum Computing." *New Scientist* 220, no. 2943 (16 November 2013): 30–31.

Cahan, David. *Helmholtz: A Life in Science.* Chicago: University of Chicago Press, 2018.

Callender, Craig. *What Makes Time Special.* New York: Oxford University Press, 2017.

Candiotto, Laura. "The Reality of Relations." *Giornale di Metafisica* 39, no. 2 (2017): 537–51.

Carleo, Giuseppe, Ignacio Cirac, Kyle Cranmer, Laurent Daudet, Maria Schuld, Naftali Tishby, Leslie Vogt-Maranto, and Lenka Zdeborová. "Machine Learning and the Physical Sciences." *Reviews of Modern Physics* 91, no. 4 (5 December 2019): 045002.

Carroll, Sean M. "The Cosmic Origins of Time's Arrow." *Scientific American* 298, no. 6 (June 2008): 48–53, 56–57.

———. "Why Boltzmann Brains Are Bad." Preprint, submitted 2 February 2017. http://arxiv.org/abs/1702.00850.

Casali, Adenauer G., Olivia Gosseries, Mario Rosanova, Mélanie Boly, Simone Sarasso, Karina R. Casali, Silvia Casarotto et al. "A Theoretically Based Index of Consciousness Independent of Sensory Processing and Behavior." *Science Translational Medicine* 5, no. 198 (14 August 2013): 198ra105.

Chalmers, David J. *The Conscious Mind: In Search of a Fundamental Theory.* New York: Oxford University Press, 1996.

———. "Consciousness and Its Place in Nature." In *The Blackwell Guide to Philosophy of Mind*, 102–42. Malden, MA: Blackwell, 2007.

———. "Could a Large Language Model Be Conscious?" Lecture, Conference on Neural Information Processing Systems, New Orleans, 28 November 2022.

———. "Dirty Secrets of Consciousness." Lecture, Foundational Questions Institute Fifth International Conference, Banff, Canada, 18 August 2016.

———. "Explaining Consciousness Scientifically: Choices and Challenges." Lecture, Science of Consciousness, Tucson, 12 April 1994.

———. "Panpsychism and Panprotopsychism." In *Consciousness in the Physical World: Perspectives on Russellian Monism*, edited by Torin Alter and Yujin Nagasawa, 246–76. New York: Oxford University Press, 2015.

———. "The Virtual and the Real." *Disputatio* 9, no. 46 (2017): 309–52.

Chalmers, David J., and Kelvin J. McQueen. "Consciousness and the Collapse of the Wave Function." In *Consciousness and Quantum Mechanics*, edited by Shan Gao, 11–63. New York: Oxford University Press, 2022.

Chisholm, Roderick M. *Human Freedom and the Self*. Lawrence: Dept. of Philosophy, University of Kansas, 1964.

Chomsky, Noam. *The Essential Chomsky*. New York: New Press, 2008.

Chrisley, Ron. "Interactive Empiricism: The Philosopher in the Machine." In *Philosophy of Engineering*, 66–71. London: Royal Academy of Engineering, 2010.

Clark, Andy, Karl J. Friston, and Sam Wilkinson. "Bayesing Qualia: Consciousness as Inference, Not Raw Datum." *Journal of Consciousness Studies* 26 (2019): 19–33.

Clay, Edmund R. *The Alternative: A Study in Psychology*. London: Macmillan, 1882.

Conselice, Christopher J., Aaron Wilkinson, Kenneth Duncan, and Alice Mortlock. "The Evolution of Galaxy Number Density at $z < 8$ and Its Implications." *Astrophysical Journal* 830, no. 2 (2016): 83.

Crease, Robert P., and James Sares. "Interview with Physicist Christopher Fuchs." *Continental Philosophy Review* 54, no. 4 (December 2021): 541–61.

Cubitt, Toby S., Jens Eisert, and Michael M. Wolf. "Extracting Dynamical Equations from Experimental Data Is NP Hard." *Physical Review Letters* 108, no. 12 (23 March 2012): 120503.

Dainton, Barry. "Temporal Consciousness." *Stanford Encyclopedia of Philosophy*, last modified 28 June 2017. https://plato.stanford.edu/entries/consciousness-temporal/.

Darrow, Karl K. "Edmond Bauer." *Physics Today* 17, no. 6 (June 1964): 86–87.

Dayan, Peter, Geoffrey E. Hinton, Radford M. Neal, and Richard S. Zemel. "The Helmholtz Machine." *Neural Computation* 7, no. 5 (September 1995): 889–904.

Dehaene, Stanislas. *Consciousness and the Brain: Deciphering How the Brain Codes Our Thoughts*. New York: Penguin, 2014.

Dempsey, Liam P. "Thinking-Matter Then and Now: The Evolution of Mind-Body Dualism." *History of Philosophy Quarterly* 26 (January 2009): 43–61.

Denchev, Vasil S., Sergio Boixo, Sergei V. Isakov, Nan Ding, Ryan Babbush, Vadim Smelyanskiy, John Martinis, and Hartmut Neven. "What Is the Computational Value of Finite-Range Tunneling?" *Physical Review X* 6, no. 3 (1 August 2016): 031015.

Dennett, Daniel C. *Consciousness Explained*. Boston: Little, Brown, 1991.

———. *Darwin's Dangerous Idea*. New York: Simon and Schuster, 1996.

———. "Expecting Ourselves to Expect: The Bayesian Brain as a Projector." *Behavioral and Brain Sciences* 36, no. 3 (June 2013): 209–10.

Descartes, René. *Meditations on First Philosophy*. Translated by Michael Moriarty. New York: Oxford University Press, 2008.

Deutsch, David. "Quantum Theory as a Universal Physical Theory." *International Journal of Theoretical Physics* 24, no. 1 (January 1985): 1–41.

DeWitt, Bryce S. "Quantum Theory of Gravity. I. The Canonical Theory." *Physical Review* 160, no. 5 (25 August 1967): 1113–48.

Donadi, Sandro, Kristian Piscicchia, Cătălina Curceanu, Lajos Diósi, Matthias Laubenstein, and Angelo Bassi. "Underground Test of Gravity-Related Wave Function Collapse." *Nature Physics* 17, no. 1 (January 2021): 74–78.

Dorr, Cian, and Frank Arntzenius. "Self-Locating Priors and Cosmological Measures." In *The Philosophy of Cosmology*, 396–428. Cambridge: Cambridge University Press, 2017.

Durham, Ian. "A Formal Model for Adaptive Free Choice in Complex Systems." *Entropy* 22, no. 5 (19 May 2020): 568.

Dvali, Gia. "Black Holes as Brains: Neural Networks with Area Law Entropy." *Fortschritte der Physik* 66, no. 4 (27 March 2018): 1800007.

Dyson, Freeman J. "Time Without End: Physics and Biology in an Open Universe." *Review of Modern Physics* 51, no. 3 (July 1979): 447–60.

Eagleman, David M. "How Does the Timing of Neural Signals Map onto the Timing of Perception?" In *Space and Time in Perception and Action*, edited by Romi Nijhawan and Beena Khurana, 216–31. Cambridge: Cambridge University Press, 2010.

Eccles, J. C. "Do Mental Events Cause Neural Events Analogously to the Probability Fields of Quantum Mechanics?" *Proceedings of the Royal Society B: Biological Sciences* 227, no. 1249 (22 May 1986): 411–28.

Einstein, Albert. *The Meaning of Relativity*. Princeton, NJ: Princeton University Press, 2005.

Elga, Adam. "Self-Locating Belief and the Sleeping Beauty Problem." *Analysis* 60, no. 2 (April 2000): 143–47.

Elkins, Katherine, and Jon Chun. "Can GPT-3 Pass a Writer's Turing Test?" *Journal of Cultural Analytics* 5, no. 2 (14 September 2020). https://culturalanalytics.org/article/17212-can-gpt-3-pass-a-writer-s-turing-test.

Everett, Hugh. "The Theory of the Universal Wave Function." In *The Many Worlds Interpretation of Quantum Mechanics*, edited by Bryce S. DeWitt and Neill Graham, 1–140. Princeton, NJ: Princeton University Press, 1973.

Faber, Roger J. *Clockwork Garden: On the Mechanistic Reduction of Living Things*. Amherst: University of Massachusetts Press, 1986.

Farhi, Edward, and Hartmut Neven. "Classification with Quantum Networks on Near Term Processors." Preprint, submitted 18 December 2017. https://arxiv.org/abs/1802.06002.

Farley, B., and W. Clark. "Simulation of Self-Organizing Systems by Digital Computer." *Transactions of the IRE Professional Group on Information Theory* 4, no. 4 (September 1954): 76–84.

Fein, Yaakov Y., Philipp Geyer, Patrick Zwick, Filip Kiałka, Sebastian Pedalino, Marcel Mayor, Stefan Gerlich, and Markus Arndt. "Quantum Superposition of Molecules Beyond 25 kDa." *Nature Physics* 15, no. 12 (23 September 2019): 1242–45.

Feynman, Richard P. *Feynman Lectures on Gravitation*. Edited by William G. Wagner and Fernando B. Morinigo. New York: Addison-Wesley, 1995.

———. "Simulating Physics with Computers." *International Journal of Theoretical Physics* 21, no. 6/7 (1982): 467–88.

Flanagan, Owen. *The Science of the Mind*. Cambridge, MA: MIT Press, 1991.

Fountas, Zafeirios, Anastasia Sylaidi, Kyriacos Nikiforou, Anil K. Seth, Murray Shanahan, and Warrick Roseboom. "A Predictive Processing Model of Episodic Memory and Time Perception." *Neural Computation* 34, no. 7 (16 June 2022): 1501–44.

Frankish, Keith. "Illusionism as a Theory of Consciousness." *Journal of Consciousness Studies* 23, no. 11–12 (2016): 11–39.

Frauchiger, Daniela, and Renato Renner. "Quantum Theory Cannot Consistently Describe the Use of Itself." *Nature Communications* 9, no. 1 (18 September 2018): 823.

Friedman, Daniel A., and Eirik Søvik. "The Ant Colony as a Test for Scientific Theories of Consciousness." *Synthese* 198, no. 2 (12 February 2019): 1457–80.

Friston, Karl J. "Hallucinations and Perceptual Inference." *Behavioral and Brain Sciences* 28, no. 6 (22 December 2005): 764–66.

———. "I Am Therefore I Think." Lecture, Foundational Questions Institute Sixth International Conference, Castelvecchio Pascoli, Italy, 23 July 2019.

———. "Learning and Inference in the Brain." *Neural Networks* 16, no. 9 (November 2003): 1325–52.

Friston, Karl J., Tamara Shiner, Thomas FitzGerald, Joseph M. Galea, Rick Adams, Harriet Brown, Raymond J. Dolan, Rosalyn Moran, Klaas Enno Stephan, and Sven Bestmann. "Dopamine, Affordance and Active Inference." *PLoS Computational Biology* 8, no. 1 (January 2012): e1002327.

Friston, Karl J., Wanja Wiese, and J. Allan Hobson. "Sentience and the Origins of Consciousness: From Cartesian Duality to Markovian Monism." *Entropy* 22, no. 5 (May 2020): 516.

Fröhlich, H. "Long-Range Coherence and Energy Storage in Biological Systems." *International Journal of Quantum Chemistry* 2, no. 5 (September 1968): 641–49.

Fuchs, Christopher A., N. David Mermin, and Rüdiger Schack. "An Introduction to QBism with an Application to the Locality of Quantum Mechanics." *American Journal of Physics* 82, no. 8 (August 2014): 749–54.

Fuchs, Christopher A., Maximilian Schlosshauer, and Blake C. Stacey. "My Struggles with the Block Universe." Preprint, submitted 10 May 2014. https://arxiv.org/abs/1405.2390.

Gan, Wen-Cong, and Fu-Wen Shu. "Holography as Deep Learning." *International Journal of Modern Physics D* 26, no. 12 (October 2017): 1743020.

Ganguli, Surya, and Haim Sompolinsky. "Compressed Sensing, Sparsity, and Dimensionality in Neuronal Information Processing and Data Analysis." *Annual Review of Neuroscience* 35, no. 1 (July 2012): 485–508.

Gavroglu, Kostas. *Fritz London: A Scientific Biography*. New York: Cambridge University Press, 1995.

Geiger, Mario, Stefano Spigler, Stéphane d'Ascoli, Levent Sagun, Marco Baity-Jesi,

Giulio Biroli, and Matthieu Wyart. "Jamming Transition as a Paradigm to Under-stand the Loss Landscape of Deep Neural Networks." *Physical Review E* 100, no. 1 (11 July 2019): 012115.

Gell-Mann, Murray, and James B. Hartle. "Quantum Mechanics in the Light of Quantum Cosmology." In *Proceedings of the 3rd International Symposium Founda-tions of Quantum Mechanics in the Light of New Technology*, edited by Shun'ichi Kobayashi, Hiroshi Ezawa, Yoshimasa Murayama, and Sadao Nomura, 321–43. Tokyo: Physical Society of Japan, 1990.

George, Mark S. "Stimulating the Brain." *Scientific American* 289 (September 2003): 66–73.

Ghirardi, G. C., A. Rimini, and T. Weber. "Unified Dynamics for Microscopic and Macroscopic Systems." *Physical Review D* 34, no. 2 (15 July 1986): 470–91.

Goff, Philip. *Galileo's Error: Foundations for a New Science of Consciousness.* New York: Pantheon, 2019.

Goodfellow, Ian J., Jean Pouget-Abadie, Mehdi Mirza, Bing Xu, David Warde-Farley, Sherjil Ozair, Aaron Courville, and Yoshua Bengio. "Generative Adversarial Net-works." Preprint, submitted 10 June 2014. https://arxiv.org/abs/1406.2661.

Goussev, Arseni, Rodolfo A. Jalabert, Horacio M. Pastawski, and Diego Wisniacki. "Loschmidt Echo." *Scholarpedia* 7, no. 8 (27 June 2012): 11687.

Greene, Brian. *The Hidden Reality: Parallel Universes and the Deep Laws of the Cosmos.* New York: Knopf, 2011.

Grossberg, Stephen. "How Does a Brain Build a Cognitive Code?" *Psychological Review* 87, no. 1 (January 1980): 1–51.

Grush, Rick. "Brain Time and Phenomenological Time." In *Cognition and the Brain*, edited by Andrew Brook and Kathleen Akins, 160–207. New York: Cambridge University Press, 2005.

Hagan, Scott, Stuart R. Hameroff, and Jack A. Tuszyński. "Quantum Computation in Brain Microtubules: Decoherence and Biological Feasibility." *Physical Review E* 65, no. 6 pt. 1 (June 2002): 061901.

Hahn, Michael, and Marco Baroni. "Tabula Nearly Rasa: Probing the Linguistic Knowledge of Character-Level Neural Language Models Trained on Unseg-mented Text." Preprint, submitted 17 June 2019. https://arxiv.org/abs/1906 .07285.

Hameroff, Stuart R. "The Quantum Origin of Life: How the Brain Evolved to Feel Good." In *On Human Nature: Biology, Psychology, Ethics, Politics, and Religion*, edited by Michel Tibayrenc and Francisco J. Ayala, 333–53. New York: Elsevier, 2017.

———. *Ultimate Computing: Biomolecular Consciousness and Nano Technology.* Amster-dam: Elsevier Science, 1987.

Hameroff, Stuart R., and Roger Penrose. "Consciousness in the Universe: An Updated Review of the 'Orch OR' Theory." In *Biophysics of Consciousness: A Foundational Approach*, edited by Roman R. Poznanski, Jack A. Tuszyński, and Todd E. Feinberg, 517–99. Singapore: World Scientific, 2016.

Hameroff, Stuart R., and Richard C. Watt. "Information Processing in Microtu-bules." *Journal of Theoretical Biology* 98, no. 4 (October 1982): 549–61.

Hardy, Quinten. "A Strange Computer Promises Great Speed." *New York Times*, 22 March 2013.

Harper, Kate. "Alexander Bain's *Mind and Body* (1872): An Underappreciated Contribution to Early Neuropsychology." *Journal of the History of the Behavioral Sciences* 55, no. 2 (April 2019): 139–60.

Hartle, James B. "The Physics of Now." *American Journal of Physics* 73, no. 2 (February 2005): 101–109.

Hartley, David. *Observations on Man, His Frame, His Duty, and His Expectations*. London: T. Tegg and Sons, 1834.

Hashimoto, Koji, Sotaro Sugishita, Akinori Tanaka, and Akio Tomiya. "Deep Learning and the AdS/CFT Correspondence." *Physical Review D* 98, no. 4 (27 August 2018): 046019.

Haun, Andrew M. "A Causal Account of Spatial Experience: IIT and the Visual Field." Lecture, Mathematical Consciousness Science, 13 August 2020.

Haun, Andrew M., and Giulio Tononi. "Why Does Space Feel the Way It Does? Towards a Principled Account of Spatial Experience." *Entropy* 21, no. 12 (December 2019): 1160.

Hausman, Daniel M. *Causal Asymmetries*. New York: Cambridge University Press, 1998.

Healey, Richard A. "Can Physics Coherently Deny the Reality of Time?" In *Time, Reality and Experience*, edited by Craig Callender, 293–316. New York: Cambridge University Press, 2002.

———. "How Many Worlds?" *Noûs* 18, no. 4 (November 1984): 591.

———. "Quantum Theory: A Pragmatist Approach." *British Journal for the Philosophy of Science* 63, no. 4 (December 2012): 729–71.

Hebb, D. O. *The Organization of Behavior: A Neuropsychological Theory*. New York: Wiley, 1949.

Heims, Steve J. *The Cybernetics Group*. Cambridge, MA: MIT Press, 1991.

Heisenberg, Werner. "The Representation of Nature in Contemporary Physics." *Daedalus* 87, no. 3 (Summer 1958): 95–108.

Heller, Michael, and Wieslaw Sasin. "Towards Noncommutative Quantization of Gravity." Preprint, submitted 1 December 1997. https://arxiv.org/abs/gr-qc/9712009.

Helmholtz, Hermann von. *Treatise on Physiological Optics*. Translated by James P. C. Southall. Rochester, NY: Optical Society of America, 1925.

Hemmo, Meir, and Orly Shenker. "Can the Past Hypothesis Explain the Psychological Arrow of Time?" In *Statistical Mechanics and Scientific Explanation*, edited by Valia Allori, 255–87. Singapore: World Scientific, 2020.

Hesse, Mary B. *Forces and Fields: The Concept of Action at a Distance in the History of Physics*. Mineola, NY: Dover, 2005.

Higgins, Irina, Nicolas Sonnerat, Loic Matthey, Arka Pal, Christopher P. Burgess, Matko Bosnjak, Murray Shanahan, Matthew Botvinick, Demis Hassabis, and Alexander Lerchner. "SCAN: Learning Hierarchical Compositional Visual Concepts." Preprint, submitted 11 July 2017. https://arxiv.org/abs/1707.03389.

Hinton, Geoffrey E. "Connectionist Learning Procedures." *Artificial Intelligence* 40, no. 1–3 (September 1989): 185–234.

Hinton, Geoffrey E., and Terrence J. Sejnowski. "Optimal Perceptual Inference." *Proceedings of the IEEE Conference on Computer Vision and Pattern Recognition* 448 (June 1983): 448–53.

Hoel, Erik P. "Causal Structure Across Scales." Lecture, Araya Brain Imaging, Tokyo, 27 April 2018.

Hoffman, Donald D. *The Case Against Reality: Why Evolution Hid the Truth from Our Eyes*. New York: W. W. Norton, 2019.

———. "The Origin of Time in Conscious Agents." *Cosmology* 18 (November 2014): 494–520. https://cosmology.com/HoffmanTime.pdf.

Hofstadter, Douglas R. "Waking Up from the Boolean Dream, or, Subcognition as Computation." In *Metamagical Themas: Questing for the Essence of Mind and Pattern*, 631–65. New York: Basic Books, 1985.

Hohwy, Jakob. *The Predictive Mind*. New York: Oxford University Press, 2013.

———. "Priors in Perception: Top-Down Modulation, Bayesian Perceptual Learning Rate, and Prediction Error Minimization." *Consciousness and Cognition* 47 (January 2017): 75–85.

Hohwy, Jakob, Bryan Paton, and Colin Palmer. "Distrusting the Present." *Phenomenology and the Cognitive Sciences* 15, no. 3 (September 2016): 315–35.

Hopfield, John J. "Neural Networks and Physical Systems with Emergent Collective Computational Abilities." *Proceedings of the National Academy of Sciences* 79, no. 8 (15 April 1982): 2554–58.

———. "Searching for Memories, Sudoku, Implicit Check Bits, and the Iterative Use of Not-Always-Correct Rapid Neural Computation." *Neural Computation* 20, no. 5 (May 2008): 1119–64.

Hopfield, John J., and David W. Tank. "'Neural' Computation of Decisions in Optimization Problems." *Biological Cybernetics* 52, no. 3 (1985): 141–52.

Howard, Don. "Revisiting the Einstein-Bohr Dialogue." *iyyun: The Jerusalem Philosophical Quarterly* 56 (January 2007): 57–90.

Hu, Huping, and Maoxin Wu. "Spin-Mediated Consciousness Theory: Possible Roles of Neural Membrane Nuclear Spin Ensembles and Paramagnetic Oxygen." *Medical Hypotheses* 63, no. 4 (2004): 633–46.

Hüttermann, Stefanie, Benjamin Noël, and Daniel Memmert. "Evaluating Erroneous Offside Calls in Soccer." *PLoS ONE* 12, no. 3 (23 March 2017): e0174358.

Idei, Hayato, Shingo Murata, Yiwen Chen, Yuichi Yamashita, Jun Tani, and Tetsuya Ogata. "A Neurorobotics Simulation of Autistic Behavior Induced by Unusual Sensory Precision." *Computational Psychiatry* 2 (2 December 2018): 164–82.

Isham, Christopher J. "Canonical Quantum Gravity and the Problem of Time." In *Integrable Systems, Quantum Groups, and Quantum Field Theories*, edited by L. A. Ibort and M. A. Rodríguez, 157–287. Dordrecht: Springer Netherlands, 1993.

Ismael, Jenann T. *How Physics Makes Us Free*. New York: Oxford University Press, 2016.

Iten, Raban, Tony Metger, Henrik Wilming, Lídia Del Rio, and Renato Renner. "Discovering Physical Concepts with Neural Networks." *Physical Review Letters* 124, no. 1 (10 January 2020): 010508.

Jackson, Frank. "Epiphenomenal Qualia." *Philosophical Quarterly* 32, no. 127 (April 1982): 127.

James, William. *The Principles of Psychology*. New York: Holt, 1890.

———. "The Spatial Quale." *Journal of Speculative Philosophy* 13, no. 1 (January 1879): 64–87.

Jammer, Max. *The Philosophy of Quantum Mechanics: The Interpretations of Quantum Mechanics in Historical Perspective*. New York: John Wiley and Sons, 1974.

Jejjala, Vishnu, Arjun Kar, and Onkar Parrikar. "Deep Learning the Hyperbolic Volume of a Knot." *Physics Letters B* 799 (10 December 2019): 135033.

Johnson, M. W., M. H. S. Amin, S. Gildert, T. Lanting, F. Hamze, N. Dickson, R. Harris et al. "Quantum Annealing with Manufactured Spins." *Nature* 473, no. 7346 (12 May 2011): 194–98.

Johnson, Steven. *Where Good Ideas Come From: The Natural History of Innovation.* New York: Penguin Group, 2010.

Joos, E., and H. D. Zeh. "The Emergence of Classical Properties Through Interaction with the Environment." *Zeitschrift für Physik B Condensed Matter* 59, no. 2 (June 1985): 223–43.

Kadowaki, Tadashi, and Hidetoshi Nishimori. "Quantum Annealing in the Transverse Ising Model." *Physical Review E* 58, no. 5 (1 November 1998): 5355–63.

Kagan, Brett J., Andy C. Kitchen, Nhi T. Tran, Forough Habibollahi, Moein Khajehnejad, Bradyn J. Parker, Anjali Bhat, Ben Rollo, Adeel Razi, and Karl J. Friston. "*In vitro* Neurons Learn and Exhibit Sentience When Embodied in a Simulated Game-World." *Neuron* 110, no. 23 (7 December 2022): 3952–69.e8.

Kaiser, David. "When Fields Collide." *Scientific American* 296, no. 6 (June 2007): 62–69.

Kanai, Ryota. "Consciousness and A.I." Lecture, Human-Level AI 2018, Prague, 25 August 2018.

———. "We Need Conscious Robots." *Nautilus* 47 (27 April 2017). https://nautil.us/we-need-conscious-robots-236579/.

Kant, Immanuel. *Critique of Pure Reason.* Edited by Vasilis Politis. London: J. M. Dent, 1993.

Károlyházy, F. "Gravitation and Quantum Mechanics of Macroscopic Objects." *Nuovo Cimento A* 42, no. 2 (March 1966): 390–402.

Kasparov, Garry. *Deep Thinking: Where Machine Intelligence Ends and Human Creativity Begins.* New York: PublicAffairs, 2017.

Kasting, James. *How to Find a Habitable Planet.* Princeton, NJ: Princeton University Press, 2012.

Keller, Helen. *The World I Live In.* New York: Century, 1908.

Kim, Hyoungkyu, and UnCheol Lee. "Criticality as a Determinant of Integrated Information Φ in Human Brain Networks." *Entropy* 21, no. 10 (October 2019): 981.

Kirchhoff, Michael D., and Tom Froese. "Where There Is Life There Is Mind: In Support of a Strong Life-Mind Continuity Thesis." *Entropy* 19, no. 4 (14 April 2017): 169.

Kitazono, Jun, Ryota Kanai, and Masafumi Oizumi. "Efficient Search for Informational Cores in Complex Systems: Application to Brain Networks." *Neural Networks* 132 (December 2020): 232–44.

Kiverstein, Julian, Mark Miller, and Erik Rietveld. "The Feeling of Grip: Novelty, Error Dynamics, and the Predictive Brain." *Synthese* 196, no. 7 (23 October 2017): 2847–69.

Kleiner, Johannes, and Erik P. Hoel. "Falsification and Consciousness." *Neuroscience of Consciousness* 2021, no. 1 (2021): niab001.

Klotz, Kelsey. "The Art of the Mistake." *The Common Reader* 11 (Summer 2019). https://commonreader.wustl.edu/c/the-art-of-the-mistake/.

Kolchinsky, Artemy, and David H. Wolpert. "Semantic Information, Autonomous

Agency and Non-Equilibrium Statistical Physics." *Interface Focus* 8, no. 6 (6 December 2018): 20180041.

Kremnizer, Kobi, and André Ranchin. "Integrated Information-Induced Quantum Collapse." *Foundations of Physics* 45, no. 8 (19 May 2015): 889–99.

Kuhn, Thomas S. *The Copernican Revolution: Planetary Astronomy in the Development of Western Thought.* Cambridge, MA: Harvard University Press, 1957.

Kuipers, Benjamin, Edward A. Feigenbaum, Peter E. Hart, and Nils J. Nilsson. "Shakey: From Conception to History." *AI Magazine* 38, no. 1 (Spring 2017): 88–103.

Lapkiewicz, Radek, Peizhe Li, Christoph Schaeff, Nathan K. Langford, Sven Ramelow, Marcin Wieśniak, and Anton Zeilinger. "Experimental Non-Classicality of an Indivisible Quantum System." *Nature* 474, no. 7352 (23 June 2011): 490–93.

Lavazza, Andrea, and Silvia Inglese. "Operationalizing and Measuring (a Kind of) Free Will (and Responsibility). Towards a New Framework for Psychology, Ethics, and Law." *Rivista Internazionale di Filosofia e Psicologia* 6, no. 1 (17 April 2015): 37–55.

LeCun, Yann. "A Theoretical Framework for Back-Propagation." In *Proceedings of the 1988 Connectionist Models Summer School,* edited by David S. Touretzky, Geoffrey E. Hinton, and Terrence J. Sejnowski, 21–28. San Mateo, CA: Morgan Kaufmann, 1988.

LeCun, Yann, Léon Bottou, Yoshua Bengio, and Patrick Haffner. "Gradient-Based Learning Applied to Document Recognition." *Proceedings of the IEEE* 86, no. 11 (November 1998): 2278–324.

LeDoux, Joseph E., Matthias Michel, and Hakwan Lau. "A Little History Goes a Long Way Toward Understanding Why We Study Consciousness the Way We Do Today." *Proceedings of the National Academy of Sciences of the United States of America* 117, no. 13 (31 March 2020): 6976–84.

Lee, Jaehoon, Yasaman Bahri, Roman Novak, Samuel S. Schoenholz, Jeffrey Pennington, and Jascha Sohl-Dickstein. "Deep Neural Networks as Gaussian Processes." Preprint, submitted 31 October 2017. https://arxiv.org/abs/1711.00165.

Lee, K. S., Y. P. Tan, L. H. Nguyen, R. P. Budoyo, K. H. Park, C. Hufnagel, Y. S. Yap et al. "Entanglement in a Qubit-qubit-tardigrade System." *New Journal of Physics* 24, no. 12 (December 2022): 123024.

Leibniz, Gottfried. *Leibniz's Monadology: A New Translation and Guide.* Translated by Lloyd Strickland. Edinburgh: Edinburgh University Press, 2014.

Levati, M. Vittoria, Matthias Uhl, and Ro'i Zultan. "Imperfect Recall and Time Inconsistencies: An Experimental Test of the Absentminded Driver 'Paradox.'" *International Journal of Game Theory* 43, no. 1 (23 April 2013): 65–88.

List, Christian. "What Is It Like to Be a Group Agent?" *Noûs* 52, no. 2 (28 July 2016): 295–319.

Little, W. A. "The Existence of Persistent States in the Brain." *Mathematical Biosciences* 19, no. 1–2 (February 1974): 101–20.

Litwin, Piotr, and Marcin Miłkowski. "Unification by Fiat: Arrested Development of Predictive Processing." *Cognitive Science* 44, no. 7 (July 2020): e12867.

Locke, John. *An Essay Concerning Humane Understanding.* London: Awnsham and John Churchill, 1706.

London, Fritz, and Edmond Bauer. "The Theory of Observation in Quantum Mechanics." In *Quantum Theory and Measurement*, edited by John Archibald Wheeler and Wojciech Hubert Zurek, 217–59. Princeton, NJ: Princeton University Press, 1983.

Loorits, Kristjan. "Qualities in the World, in Science, and in Consciousness." *Journal of Consciousness Studies* 29, no. 11 (2022): 108–30.

———. "Structural Qualia: A Solution to the Hard Problem of Consciousness." *Frontiers in Psychology* 5, no. e1000462 (18 March 2014): 237.

Lundholm, Ida V., Helena Rodilla, Weixiao Y. Wahlgren, Annette Duelli, Gleb Bourenkov, Josip Vukusic, Ran Friedman, Jan Stake, Thomas Schneider, and Gergely Katona. "Terahertz Radiation Induces Non-Thermal Structural Changes Associated with Fröhlich Condensation in a Protein Crystal." *Structural Dynamics* 2, no. 5 (September 2015): 054702.

Maccone, Lorenzo. "Quantum Solution to the Arrow-of-Time Dilemma." *Physical Review Letters* 103, no. 8 (21 August 2009): 080401.

MacKay, Donald M. "The Epistemological Problem for Automata." In *Automata Studies*, edited by Claude E. Shannon and J. McCarthy, 235–52. Princeton, NJ: Princeton University Press, 1956.

———. "Mindlike Behaviour in Artefacts." *British Journal for the Philosophy of Science* 2, no. 6 (October 1951): 105–21.

Madau, Piero, and Mark Dickinson. "Cosmic Star-Formation History." *Annual Review of Astronomy and Astrophysics* 52, no. 1 (August 2014): 415–86.

Malicki, Jarema J., and Colin A Johnson. "The Cilium: Cellular Antenna and Central Processing Unit." *Trends in Cell Biology* 27, no. 2 (February 2017): 126–40.

Marcus, Gary, and Ernest Davis. "Are Neural Networks About to Reinvent Physics?" *Nautilus* 78 (21 November 2019). https://nautil.us/are-neural-networks-about-to-reinvent-physics-237619/.

Marletto, C., D. M. Coles, T. Farrow, and V. Vedral. "Entanglement Between Living Bacteria and Quantized Light Witnessed by Rabi Splitting." *Journal of Physics Communications* 2, no. 10 (10 October 2018): 101001.

Marshall, I. N. "Consciousness and Bose-Einstein Condensates." *New Ideas in Psychology* 7, no. 1 (1989): 73–83.

Marstaller, Lars, Arend Hintze, and Christoph Adami. "The Evolution of Representation in Simple Cognitive Networks." *Neural Computation* 25, no. 8 (August 2013): 2079–107.

Maudlin, Tim. *Quantum Non-Locality and Relativity: Metaphysical Intimations of Modern Physics*, 2nd ed. Malden, MA: Blackwell, 2002.

McFadden, Johnjoe, and Jim Al-Khalili. *Life on the Edge: The Coming of Age of Quantum Biology*. New York: Crown, 2014.

McGinn, Colin. "Consciousness and Space." *Journal of Consciousness Studies* 2, no. 3 (1 March 1995): 220–30.

Mediano, Pedro A. M., Fernando E. Rosas, Juan Carlos Farah, Murray Shanahan, Daniel Bor, and Adam B. Barrett. "Integrated Information as a Common Signature of Dynamical and Information-Processing Complexity." *Chaos* 32, no. 1 (January 2022): 013115.

Mehrabi, Ninareh, Fred Morstatter, Nripsuta Saxena, Kristina Lerman, and Aram Galstyan. "A Survey on Bias and Fairness in Machine Learning." *ACM Computing Surveys* 54, no. 6 (July 2021): 1–35.

Mehta, Pankaj, and David J. Schwab. "An Exact Mapping Between the Variational Renormalization Group and Deep Learning." Preprint, submitted 14 October 2014. https://arxiv.org/abs/1410.3831.

Miller, Kristie, Alex Holcombe, and Andrew James Latham. "Temporal Phenomenology: Phenomenological Illusion Versus Cognitive Error." *Synthese* 197, no. 2 (February 2020): 751–71.

Mlodinow, Leonard, and Todd A. Brun. "Relation Between the Psychological and Thermodynamic Arrows of Time." *Physical Review E* 89, no. 5 (May 2014): 052102.

Mora, Thierry, and William Bialek. "Are Biological Systems Poised at Criticality?" *Journal of Statistical Physics* 144, no. 2 (2 June 2011): 268–302.

Morange, Michel. *The Black Box of Biology: A History of the Molecular Revolution.* Cambridge, MA: Harvard University Press, 2020.

Mørch, Hedda Hassel. "Is the Integrated Information Theory of Consciousness Compatible with Russellian Panpsychism?" *Erkenntnis* 84, no. 5 (10 April 2018): 1065–85.

Moreva, Ekaterina, Giorgio Brida, Marco Gramegna, Vittorio Giovannetti, Lorenzo Maccone, and Marco Genovese. "Time from Quantum Entanglement: An Experimental Illustration." *Physical Review A* 89, no. 5 (20 May 2014): 052122.

Moser, Edvard I., Emilio Kropff, and May-Britt Moser. "Place Cells, Grid Cells, and the Brain's Spatial Representation System." *Annual Review of Neuroscience* 31 (2008): 69–89.

Mott, Alex, Joshua Job, Jean-Roch Vlimant, Daniel Lidar, and Maria Spiropulu. "Solving a Higgs Optimization Problem with Quantum Annealing for Machine Learning." *Nature* 550, no. 7676 (19 October 2017): 375–79.

Muciño, Ricardo, Elias Okon, and Daniel Sudarsky. "Assessing Relational Quantum Mechanics." *Synthese* 200, no. 5 (October 2022): 399.

Müller, Markus P. "Law Without Law: From Observer States to Physics via Algorithmic Information Theory." *Quantum* 4 (20 July 2020): 301.

Muñoz, Roberto N., Angus Leung, Aidan Zecevik, Felix A. Pollock, Dror Cohen, Bruno van Swinderen, Naotsugu Tsuchiya, and Kavan Modi. "General Anesthesia Reduces Complexity and Temporal Asymmetry of the Informational Structures Derived from Neural Recordings in *Drosophila*." *Physical Review Research* 2 (22 May 2020): 023219.

Muñoz Garganté, Núria. "A Physicist's Road to Emergence: Revisiting the Story of 'More Is Different.'" Lecture, Max Planck Institute for the History of Science, Berlin, 3 September 2019.

Musser, George. "Build Your Own Artificial Neural Network. It's Easy!" *Nautilus* (20 September 2020). https://nautil.us/build-your-own-artificial-neural-network-its-easy-237976/.

———. "How Autism May Stem from Problems with Prediction." *Spectrum News* (7 March 2018). https://www.spectrumnews.org/features/deep-dive/autism-may-stem-problems-prediction/.

———. "Schrödinger's A.I. Could Test the Foundations of Reality." *FQxI Blogs*, 19 September 2022. https://fqxi.org/community/articles/display/266.

———. "Watching the Watchmen: Demystifying the Frauchiger-Renner Experiment—Musings from Lidia del Rio and More at the 6th FQxI Meeting." *FQxI Blogs*, 24 December 2019. https://fqxi.org/community/forum/topic/3354.

Nautrup, Hendrik Poulsen, Tony Metger, Raban Iten, Sofiene Jerbi, Lea M. Trenkwalder, Henrik Wilming, Hans J. Briegel, and Renato Renner. "Operationally Meaningful Representations of Physical Systems in Neural Networks." *Machine Learning: Science and Technology* 3, no. 4 (December 2022): 045025.

Neal, Radford M. *Bayesian Learning for Neural Networks: Lecture Notes in Statistics.* New York: Springer New York, 1996.

Neven, Hartmut. "Car Detector Trained with the Quantum Adiabatic Algorithm." Demonstration, Conference on Neural Information Processing Systems, Vancouver, Canada, 8 December 2009.

Newton, Isaac. *Opticks: or, A Treatise of the Reflections, Refractions, Inflections and Colours of Light.* London: Royal Society, 1718.

Nishimori, Hidetoshi, and Yoshihiko Nonomura. "Quantum Effects in Neural Networks." *Journal of the Physical Society of Japan* 65, no. 12 (15 December 1996): 3780–96.

Nye, Maxwell, Anders Johan Andreassen, Guy Gur-Ari, Henryk Michalewski, Jacob Austin, David Bieber, David Dohan et al. "Show Your Work: Scratchpads for Intermediate Computation with Language Models." Preprint, submitted 30 November 2021. https://arxiv.org/pdf/2112.00114.

O'Connell, A. D., M. Hofheinz, M. Ansmann, Radoslaw C. Bialczak, M. Lenander, Erik Lucero, M. Neeley et al. "Quantum Ground State and Single-Phonon Control of a Mechanical Resonator." *Nature* 464, no. 7289 (1 April 2010): 697–703.

Oddie, Graham. "Armstrong on the Eleatic Principle and Abstract Entities." *Philosophical Studies* 41, no. 2 (March 1982): 285–95.

Odegaard, Brian, Robert T. Knight, and Hakwan Lau. "Should a Few Null Findings Falsify Prefrontal Theories of Conscious Perception?" *Journal of Neuroscience* 37, no. 40 (4 October 2017): 9593–602.

Ogden, Ruth S. "The Passage of Time During the UK Covid-19 Lockdown." *PLoS ONE* 15, no. 7 (6 July 2020): e0235871.

Okon, Elias, and Miguel Ángel Sebastián. "A Consciousness-Based Quantum Objective Collapse Model." *Synthese* 197, no. 9 (27 July 2018): 3947–67.

Olah, Chris, Alexander Mordvintsev, and Ludwig Schubert. "Feature Visualization." *Distill*, 7 November 2017. https://distill.pub/2017/feature-visualization/.

Olmsted, J. B., and G. G. Borisy. "Microtubules." *Annual Review of Biochemistry* 42 (1973): 507–40.

Oriti, Daniele. "Levels of Spacetime Emergence in Quantum Gravity." In *Philosophy Beyond Spacetime*, edited by Christian Wüthrich, Baptiste Le Bihan, and Nick Huggett, 16–40. New York: Oxford University Press, 2021.

Page, Don N., and William K. Wootters. "Evolution Without Evolution: Dynamics Described by Stationary Observables." *Physical Review D* 27, no. 12 (15 June 1983): 2885–92.

Päs, Heinrich. *The One: How an Ancient Idea Holds the Future of Physics.* New York: Basic Books, 2023.

Pearl, Judea. *Causality*, 2nd ed. New York: Cambridge University Press, 2009.

Pearl, Judea, and Dana Mackenzie. *The Book of Why: The New Science of Cause and Effect.* New York: Basic Books, 2018.

Penrose, Roger. *The Emperor's New Mind: Concerning Computers, Minds, and the Laws of Physics*. New York: Penguin Books, 1991.

———. "Gravity and State Vector Reduction." In *Quantum Concepts in Space and Time*, edited by Roger Penrose and Christopher J. Isham, 129–46. New York: Oxford University Press, 1986.

———. *Shadows of the Mind: A Search for the Missing Science of Consciousness*. New York: Oxford University Press, 1994.

Pereboom, Derk. *Consciousness and the Prospects of Physicalism*. New York: Oxford University Press, 2011.

Peres, Asher. "Unperformed Experiments Have No Results." *American Journal of Physics* 46, no. 7 (July 1978): 745–47.

Piccione, Michele, and Ariel Rubinstein. "On the Interpretation of Decision Problems with Imperfect Recall." *Games and Economic Behavior* 20, no. 1 (July 1997): 3–24.

Pienaar, Jacques L. "A Quintet of Quandaries: Five No-Go Theorems for Relational Quantum Mechanics." *Foundations of Physics* 51 (4 October 2021): 97.

Pinker, Steven. *How the Mind Works*. New York: Penguin Random House, 2015.

Plato. "Parmenides." In *Readings in Ancient Greek Philosophy: From Thales to Aristotle*, edited by S. Marc Cohen, Patricia Curd, and C. D. C. Reeve, 432–41. Cambridge, MA: Hackett, 1995.

Polchinski, Joseph. *String Theory: An Introduction to the Bosonic String*. New York: Cambridge University Press, 1998.

Price, Huw. "Causal Perspectivalism." In *Causation, Physics, and the Constitution of Reality*, edited by Huw Price and Richard Corry, 250–92. New York: Oxford University Press, 2007.

———. *Time's Arrow and Archimedes' Point: New Directions for the Physics of Time*. New York: Oxford University Press, 1996.

Principe, Lawrence M. "Reflections on Newton's Alchemy in Light of the New Historiography of Alchemy." In *Newton and Newtonianism*, edited by James E. Force and Sarah Hutton, 205–19. Boston: Kluwer Academic, 2004.

Proietti, Massimiliano, Alexander Pickston, Francesco Graffitti, Peter Barrow, Dmytro Kundys, Cyril Branciard, Martin Ringbauer, and Alessandro Fedrizzi. "Experimental Test of Local Observer Independence." *Science Advances* 5, no. 9 (20 September 2019): eaaw9832.

Rahimi, Ali. "Back When We Were Young." Lecture, Conference on Neural Information Processing Systems, Long Beach, CA, 5 December 2017.

Rao, Rajesh P. N., and Dana H. Ballard. "Predictive Coding in the Visual Cortex: A Functional Interpretation of Some Extra-Classical Receptive-Field Effects." *Nature Neuroscience* 2, no. 1 (January 1999): 79–87.

Rasmussen, Carl Edward. "Gaussian Processes in Machine Learning." In *Advanced Lectures on Machine Learning: Lecture Notes in Computer Science*, edited by Olivier Bousquet, Ulrike von Luxburg, and Gunnar Rätsch, 63–71. Berlin, Heidelberg: Springer Berlin Heidelberg, 2004.

Rauch, D., J. Handsteiner, A. Hochrainer, J. Gallicchio, A. S. Friedman, C. Leung, B. Liu et al. "Cosmic Bell Test Using Random Measurement Settings from High-Redshift Quasars." *Physical Review Letters* 121, no. 8 (24 August 2018): 080403.

Roose, Kevin. "Bing's Chatbot Drew Me In and Creeped Me Out." *New York Times*, 17 February 2023.

Rovelli, Carlo. "Consciousness, Time, Quantum." Lecture, The Science of Consciousness, 15 September 2020.

———. "Forget Time." Preprint, submitted 23 March 2009. https://arxiv.org/abs/0903.3832.

———. *Helgoland*. New York: Penguin, 2021.

———. "Is Time's Arrow Perspectival?" In *The Philosophy of Cosmology*, edited by Khalil Chamcham, Joseph Silk, John D. Barrow, and Simon Saunders, 285–96. New York: Cambridge University Press, 2017.

———. *Quantum Gravity*. New York: Cambridge University Press, 2004.

———. "Relational Quantum Mechanics." *International Journal of Theoretical Physics* 35, no. 8 (August 1996): 1637–78.

———. *What Is Time? What Is Space?* Translated by J. C. van den Berg. Rome: Di Renzo Editore, 2006.

Rubin, Sergio, Thomas Parr, Lancelot Da Costa, and Karl Friston. "Future Climates: Markov Blankets and Active Inference in the Biosphere." *Journal of the Royal Society Interface* 17, no. 172 (November 2020): 20200503.

Rudy, Samuel H., Steven L. Brunton, Joshua L. Proctor, and J. Nathan Kutz. "Data-Driven Discovery of Partial Differential Equations." *Science Advances* 3, no. 4 (April 2017): e1602614.

Rumelhart, David E., Geoffrey E. Hinton, and Ronald J. Williams. "Learning Representations by Back-Propagating Errors." *Nature* 323, no. 6088 (9 October 1986): 533–36.

Russell, Bertrand. "On the Notion of Cause." *Proceedings of the Aristotelian Society* 13 (1912): 1–26.

Sacks, Oliver. *The Man Who Mistook His Wife for a Hat: And Other Clinical Tales*. New York: Simon and Schuster, 1998.

Salisbury, Donald. "Toward a Quantum Theory of Gravity: Syracuse 1949–1962." In *The Renaissance of General Relativity in Context*, edited by Alexander S. Blum, Roberto Lalli, and Jürgen Renn, 221–55. Cham, Switzerland: Springer International, 2020.

Saysani, Armin, Michael C. Corballis, and Paul M. Corballis. "Colour Envisioned: Concepts of Colour in the Blind and Sighted." *Visual Cognition* 26, no. 5 (May 2018): 382–92.

Sayood, Khalid. *Introduction to Data Compression*, 4th ed. Waltham, MA: Morgan Kaufmann, 2012.

Schlosshauer, Maximilian, and Arthur Fine. "Decoherence and the Foundations of Quantum Mechanics." In *The Frontiers Collection: Quantum Mechanics at the Crossroads*, edited by James Evans and Alan S. Thorndike, 125–48. Heidelberg: Springer Berlin Heidelberg, 2007.

Schmidt, M., and H. Lipson. "Distilling Free-Form Natural Laws from Experimental Data." *Science* 324, no. 5923 (3 April 2009): 81–85.

Schneider, Susan. *Artificial You*. Princeton, NJ: Princeton University Press, 2019.

Schoenbrun, David. *Soldiers of the Night: The Story of the French Resistance*. New York: Dutton, 1980.

Schoenholz, Samuel S., Justin Gilmer, Surya Ganguli, and Jascha Sohl-Dickstein.

"Deep Information Propagation." Preprint, submitted 4 November 2016. https://arxiv.org/abs/1611.01232.

Schramski, Sam. "Running Is Always Blind." *Nautilus* 38 (30 June 2016). https://nautil.us/running-is-always-blind-236003/.

Schrödinger, Erwin. *Nature and the Greeks*. Cambridge: Cambridge University Press, 1954.

Schuld, Maria, Ilya Sinayskiy, and Francesco Petruccione. "Quantum Computing for Pattern Classification." In *Lecture Notes in Computer Science: PRICAI 2014: Trends in Artificial Intelligence*, edited by Duc-Nghia Pham and Seong-Bae Park, 208–20. Cham, Switzerland: Springer International, 2014.

Schweber, Silvan S. "Physics, Community and the Crisis in Physical Theory." *Physics Today* 46, no. 11 (November 1993): 34–40.

Scoles, Sarah. "NASA's Space Crash Succeeded in Forcing Asteroid onto New Path." *New York Times*, 12 October 2022.

Seager, William. *Natural Fabrications: Science, Emergence and Consciousness*. New York: Springer, 2012.

———. "Neutral Monism and the Scientific Study of Consciousness." Lecture, Mathematical Consciousness Science, 15 December 2020.

Sebens, Charles T., and Sean M. Carroll. "Self-Locating Uncertainty and the Origin of Probability in Everettian Quantum Mechanics." *British Journal for the Philosophy of Science* 69, no. 1 (March 2018): 25–74.

Seiberg, Nathan. "Emergent Spacetime." In *The Quantum Structure of Space and Time*, edited by David J. Gross, Marc Henneaux, and Alexander Sevrin, 162–213. Hackensack, NJ: World Scientific, 2007.

Seif, Alireza, Mohammad Hafezi, and Christopher Jarzynski. "Machine Learning the Thermodynamic Arrow of Time." *Nature Physics* 17, no. 1 (January 2021): 105–13.

Sejnowski, Terrence J. "The Unreasonable Effectiveness of Deep Learning in Artificial Intelligence." *Proceedings of the National Academy of Sciences* 117, no. 48 (1 December 2020): 30033–38.

Sejnowski, Terrence J., and Charles R. Rosenberg. "Parallel Networks That Learn to Pronounce English Text." *Complex Systems* 1, no. 1 (February 1987): 145–68.

Selman, Bart. "Dirty Secrets of Artificial Intelligence." Lecture, Foundational Questions Institute Fifth International Conference, Banff, Canada, 18 August 2016.

Seth, Anil K. *Being You: A New Science of Consciousness*. New York: Penguin Random House, 2021.

Sharp, Kim, and Franz Matschinsky. "Translation of Ludwig Boltzmann's Paper 'On the Relationship Between the Second Fundamental Theorem of the Mechanical Theory of Heat and Probability Calculations Regarding the Conditions for Thermal Equilibrium.'" *Entropy* 17, no. 4 (April 2015): 1971–2009.

Shimony, Abner. "Role of the Observer in Quantum Theory." *American Journal of Physics* 31, no. 10 (October 1963): 755–73.

Silver, David, Aja Huang, Chris J. Maddison, Arthur Guez, Laurent Sifre, George van den Driessche, Julian Schrittwieser et al. "Mastering the Game of Go with Deep Neural Networks and Tree Search." *Nature* 529, no. 7587 (28 January 2016): 484–89.

Smith, Jordan, Hadi Zadeh Haghighi, Dennis Salahub, and Christoph Simon. "Radical Pairs May Play a Role in Xenon-Induced General Anesthesia." *Scientific Reports* 11, no. 1 (18 March 2021): 6287.

Smith, Robert W. "Beyond the Galaxy: The Development of Extragalactic Astronomy 1885–1965, Part 1." *Journal for the History of Astronomy* 39, no. 1 (February 2008): 91–119.

Smolin, Lee. *Time Reborn: From the Crisis in Physics to the Future of the Universe.* New York: Houghton Mifflin Harcourt, 2013.

Solomonoff, R. J. "A Formal Theory of Inductive Inference. Part I." *Information and Control* 7, no. 1 (March 1964): 1–22.

Spalding, Kirsty L., Ratan D. Bhardwaj, Bruce A. Buchholz, Henrik Druid, and Jonas Frisén. "Retrospective Birth Dating of Cells in Humans." *Cell* 122, no. 1 (15 July 2005): 133–43.

Srinivasan, M. V., S. B. Laughlin, and A. Dubs. "Predictive Coding: A Fresh View of Inhibition in the Retina." *Proceedings of the Royal Society B: Biological Sciences* 216, no. 1205 (22 November 1982): 427–59.

Stachel, John. "The Other Einstein: Einstein Contra Field Theory." *Science in Context* 6, no. 1 (Spring 1993): 275–90.

Stalzer, Mark, and Chao Ju. "Automated Rediscovery of the Maxwell Equations." *Applied Sciences* 9, no. 14 (19 July 2019): 2899.

Stanley, Richard P. "Qualia Space." *Journal of Consciousness Studies* 6, no. 1 (1 January 1999): 49–60.

Stapp, Henry P. "Quantum Propensities and the Brain-Mind Connection." *Foundations of Physics* 21, no. 12 (November 1991): 1451–77.

Steinhardt, Paul J. "The Inflation Debate." *Scientific American* 304, no. 4 (April 2011): 36–43.

Steinmetz, Ralf. "Human Perception of Jitter and Media Synchronization." *IEEE Journal on Selected Areas in Communications* 14, no. 1 (January 1996): 61–72.

Sutton, John. *Philosophy and Memory Traces.* New York: Cambridge University Press, 1998.

Swazey, Judith P. "Forging a Neuroscience Community: A Brief History of the Neurosciences Research Program." In *The Neurosciences: Paths of Discovery*, edited by George Adelman, Judith P. Swazey, Frederic G. Worden, and Francis Otto Schmitt, 529–46. Cambridge, MA: MIT Press, 1975.

Tani, Jun. "An Interpretation of the 'Self' from the Dynamical Systems Perspective: A Constructivist Approach." *Journal of Consciousness Studies* 5 (1 May 1998): 516–42.

———. "Learning to Generate Articulated Behavior Through the Bottom-Up and the Top-Down Interaction Processes." *Neural Networks* 16, no. 1 (January 2003): 11–23.

———. "Model-Based Learning for Mobile Robot Navigation from the Dynamical Systems Perspective." *IEEE Transactions on Systems, Man, and Cybernetics, Part B (Cybernetics)* 26, no. 3 (June 1996): 421–36.

Tao, Terence. "The Erdős Discrepancy Problem." *Discrete Analysis* 1 (28 February 2016). https://discreteanalysisjournal.com/article/609.

Taranalli, Veeresh, Hironori Uchikawa, and Paul H. Siegel. "Channel Models for Multi-Level Cell Flash Memories Based on Empirical Error Analysis." *IEEE Transactions on Communications* 64, no. 8 (August 2016): 3169–81.

Tegmark, Max. "Consciousness as a State of Matter." *Chaos, Solitons & Fractals* 76 (July 2015): 238–70.

———."Importance of Quantum Decoherence in Brain Processes." *Physical Review E* 61, no. 4 (April 2000): 4194–206.

———. "Improved Measures of Integrated Information." *PLoS Computational Biology* 12, no. 11 (21 November 2016): e1005123.

———. "Parallel Universes." *Scientific American* 288, no. 5 (May 2003): 40–51.

Tinsley, J. N., M. I. Molodtsov, R. Prevedel, D. Wartmann, J. Espigulé-Pons, M. Lauwers, and A. Vaziri. "Direct Detection of a Single Photon by Humans." *Nature Communications* 7 (19 July 2016): 12172.

Tononi, Giulio, Larissa Albantakis, Melanie Boly, Chiara Cirelli, and Christof Koch. "Only What Exists Can Cause: An Intrinsic View of Free Will." Preprint, submitted 4 June 2022. https://arxiv.org/abs/2206.02069.

Tononi, Giulio, Melanie Boly, Marcello Massimini, and Christof Koch. "Integrated Information Theory: From Consciousness to Its Physical Substrate." *Nature Reviews Neuroscience* 17, no. 7 (July 2016): 450–61.

Tononi, Giulio, and Christof Koch. "Consciousness: Here, There and Everywhere?" *Philosophical Transactions of the Royal Society B: Biological Sciences* 370, no. 1668 (19 May 2015).

Tsuchiya, Naotsugu, Shigeru Taguchi, and Hayato Saigo. "Using Category Theory to Assess the Relationship Between Consciousness and Integrated Information Theory." *Neuroscience Research* 107 (June 2016): 1–7.

Tversky, Barbara. *Mind in Motion*. New York: Basic Books, 2019.

Udrescu, Silviu-Marian, and Max Tegmark. "AI Feynman: A Physics-Inspired Method for Symbolic Regression." *Science Advances* 6, no. 16 (15 April 2020): eaay2631.

Uvarov, A. V., A. S. Kardashin, and Jacob D. Biamonte. "Machine Learning Phase Transitions with a Quantum Processor." *Physical Review A* 102, no. 1 (15 July 2020): 012415.

Vanchurin, Vitaly. "The World as a Neural Network." *Entropy* 22, no. 11 (26 October 2020): 1210.

Van Raamsdonk, Mark. "Building Up Spacetime with Quantum Entanglement." *General Relativity and Gravitation* 42, no. 10 (19 June 2010): 2323–29.

Vardanyan, Mihran, Roberto Trotta, and Joseph Silk. "Applications of Bayesian Model Averaging to the Curvature and Size of the Universe." *Monthly Notices of the Royal Astronomical Society: Letters* 413, no. 1 (May 2011): L91–95.

Vaswani, Ashish, Noam Shazeer, Niki Parmar, Jakob Uszkoreit, Llion Jones, Aidan N. Gomez, Lukasz Kaiser, and Illia Polosukhin. "Attention Is All You Need." Preprint, submitted 12 June 2017. https://arxiv.org/abs/1706.03762.

von Neumann, John. "The General and Logical Theory of Automata." In *Cerebral Mechanisms in Behavior: The Hixon Symposium*, edited by Lloyd A. Jeffress, 1–42. New York: Hafner, 1967.

———. *Mathematical Foundations of Quantum Mechanics*. Translated by Robert T. Beyer. Princeton, NJ: Princeton University Press, 1955.

von Weizsäcker, C. F. "Physics and Philosophy." In *The Physicist's Conception of Nature*, edited by Jagdish Mehra, 736–46. Dordrecht: Springer Netherlands, 1973.

Wallace, David. "Everett and Structure." *Studies in History and Philosophy of Science*

Part B: Studies in History and Philosophy of Modern Physics 34, no. 1 (March 2003): 87–105.

Warren, Howard Crosby. *A History of the Association Psychology.* New York: Charles Scribner's Sons, 1921.

Watson, Richard A., and Eörs Szathmáry. "How Can Evolution Learn?" *Trends in Ecology and Evolution* 31, no. 2 (February 2016): 147–57.

Wharton, Ken B. "Time-Symmetric Boundary Conditions and Quantum Foundations." *Symmetry* 2, no. 1 (March 2010): 272–83.

Wheeler, John Archibald. "Genesis and Observership." In *Foundational Problems in the Special Sciences*, edited by Robert E. Butts and Jaakko Hintikka, 3–33. Dordrecht: Springer Netherlands, 1977.

———. Papers 52:140. Relativity Notebook #14, 27 January 1967. American Philosophical Society Library, Philadelphia.

———. "Pregeometry: Motivations and Prospects." In *Quantum Theory and Gravitation*, edited by A. R. Marlow, 1–11. New York: Academic Press, 1980.

Wiese, Wanja, and Karl J. Friston. "The Neural Correlates of Consciousness Under the Free Energy Principle: From Computational Correlates to Computational Explanation." *Philosophy and the Mind Sciences* 2 (22 September 2021).

Wigner, Eugene Paul. "Remarks on the Mind-Body Question." In *The Scientist Speculates: An Anthology of Partly-Baked Ideas*, edited by Irving John Good, 284–302. London: Heinemann, 1962.

———. "Review of the Quantum Mechanical Measurement Problem." In *Quantum Optics, Experimental Gravity, and Measurement Theory*, edited by Pierre Meystre and Marlan O. Scully, 43–63. Boston: Springer U.S., 1983.

Wright, Sewall. "Correlation and Causation." *Journal of Agricultural Research* 20, no. 7 (3 January 1921): 557–85.

Yaida, Sho. "Non-Gaussian Processes and Neural Networks at Finite Widths." Preprint, submitted 30 September 2019. https://arxiv.org/abs/1910.00019.

Yamashita, Yuichi, and Jun Tani. "Spontaneous Prediction Error Generation in Schizophrenia." *PLoS ONE* 7, no. 5 (30 May 2012): e37843.

Zhang, Chiyuan, Samy Bengio, Moritz Hardt, Benjamin Recht, and Oriol Vinyals. "Understanding Deep Learning Requires Rethinking Generalization." Preprint, submitted 10 November 2016. https://arxiv.org/abs/1611.03530.

INTERVIEWS AND PERSONAL COMMUNICATION

Scott Aaronson. 18 September 2011, 2 November 2022
Emily Adlam. 7 November 2022
Anthony Aguirre. 22 January 2014
Larissa Albantakis. 22 July 2019, 2 December 2019, 2 April 2020, 5 October 2020, 20 October 2021, 22 October 2021
David Albert. 28 March 2011, 11 April 2011, 24 April 2017, 3 May 2021
Igor Aleksander. 23 July 2018
Holly Andersen. 28 October 2020
Joscha Bach. 6 June 2017, 4 October 2017, 16 October 2017, 24 August 2018
Yasaman Bahri. 4 September 2019, 29 December 2019, 13 January 2020
Vijay Balasubramanian. 4 December 2018, 23 September 2019

Elizabeth Behrman. 15 December 2017, 21 January 2021, 22 March 2021
Mani Bhaumik. 27 June 2019, 29 June 2020
Raphael Bousso. 17 February 2011, 2 November 2022
Vern Brownell. 20 December 2018
Časlav Brukner. 4 August 2020, 6 February 2022
Craig Callender. 28 May 2002, 30 June 2020
Eric Cavalcanti. 28 July 2020, 29 July 2020, 17 August 2020, 8 February 2022
Ron Chrisley. 6 January 2016, 28 October 2017
Andy Clark. 8 November 2017, 15 February 2018
Kyle Cranmer. 14 December 2018, 4 October 2021
Cătălina Curceanu. 4 September 2020, 21 April 2021
Daniel Dennett. 3 April 2015
Dennis Dieks. 11 August 2020, 14 August 2020, 15 August 2020, 1 October 2020, 7 October 2020, 9 May 2021, 12 May 2021
Robbert Dijkgraaf. 7 December 2021
Gia Dvali. 13 March 2018, 19 March 2018, 31 March 2018, 2 April 2018, 29 January 2021
George Ellis. 24 March 2017, 25 March 2017
Zafeirios Fountas. 2 October 2020
Karl Friston. 24 October 2017, 25 January 2018, 27 January 2018, 28 September 2018, 23 July 2019
Christopher Fuchs. 7 July 2004, 5 May 2005, 7 April 2015, 3 July 2019, 8 November 2019, 11 June 2021
Surya Ganguli. 28 August 2019
Philip Goff. 15 July 2021, 28 July 2021, 1 August 2021, 10 August 2021
Rick Grush. 30 August 2021
Mile Gu. 29 September 2018
Nicholas Guttenberg. 14 October 2017, 17 December 2018, 18 February 2023
Joseph Halpern. 19 April 2017
Stuart Hameroff. 28 February 2020, 29 February 2020, 1 March 2020, 2 March 2020, 4 March 2020
Koji Hashimoto. 7 May 2020, 22 September 2021
Andrew Haun. 20 August 2020, 21 October 2021
Elad Hazan. 20 February 2019
Richard Healey. 21 November 2011, 28 June 2017, 26 May 2019, 8 May 2021
Irina Higgins. 25 August 2018, 22 July 2021
Erik Hoel. 1 June 2016, 28 March 2017, 31 March 2017, 3 April 2017, 5 April 2017
Donald Hoffman. 6 July 2020
John Hopfield. 28 November 2018, 11 December 2018, 8 January 2019, 15 January 2019, 31 January 2019, 22 April 2019, 13 December 2020, 23 February 2023
Jenann Ismael. 29 October 2013, 5 April 2017
Mark Johnson. 8 October 2020
Bjørn Erik Juel. 17 February 2021
Yann LeCun. 4 June 2019, 19 November 2019, 21 November 2019, 9 December 2021
Christian List. 9 March 2015, 13 March 2015, 31 March 2015, 12 April 2015, 27 May 2015, 30 April 2019, 2 May 2019, 3 May 2019

Seth Lloyd. 19 May 2007, 6 January 2014, 7 December 2017

Kristjan Loorits. 31 July 2020, 6 August 2020, 11 August 2020, 7 September 2022, 15 September 2022

Kelvin McQueen. 21 March 2017, 22 March 2017, 29 September 2018, 27 September 2020, 28 September 2020

Marc Mézard. 22 November 2019

Markus Müller. 29 September 2018, 26 October 2018, 24 October 2019, 29 May 2021, 31 May 2021

Hidetoshi Nishimori. 13 October 2017, 11 May 2020, 21 January 2021

Heinrich Päs. 26 June 2019, 28 June 2019

Jeffrey Pennington. 8 July 2019

Roger Penrose. 15 March 2010, 6 April 2018, 4 September 2020

Jacques Pienaar. 29 June 2021, 1 July 2021, 6 July 2021, 7 July 2021, 26 July 2021

Joseph Polchinski. 9 January 2009

Sandu Popescu. 17 November 2010

Huw Price. 22 July 2011, 25 October 2013

Giovanni Rabuffo. 11 November 2020, 13 October 2022, 20 October 2022

Renato Renner. 2 November 2019, 16 November 2019, 5 August 2020, 3 June 2022

Daniel Roberts. 20 November 2019, 24 March 2020, 21 April 2020, 13 January 2021, 27 September 2021, 3 April 2023

Warrick Roseboom. 2 October 2020

Carlo Rovelli. 20 January 2007, 28 January 2007, 27 May 2007, 28 May 2007, 9 March 2008, 9 October 2008, 9 January 2014, 26 June 2020

Jürgen Schmidhuber. 30 June 2019

Susan Schneider. 25 May 2017, 26 June 2018

Maria Schuld. 1 November 2017, 19 November 2019, 21 November 2019

Terry Sejnowski. 14 November 2019, 15 November 2019

Anil Seth. 9 November 2017, 18 July 2018, 25 February 2021

Lee Smolin. 25 August 2021

Jascha Sohl-Dickstein. 27 August 2019, 29 January 2021

Haim Sompolinsky. 19 November 2019

Joanna Szczotka. 20 August 2020

Jun Tani. 16 October 2017, 22 October 2017, 29 October 2017, 16 February 2019

Max Tegmark. 25 March 2015, 23 March 2017, 21 July 2019, 23 July 2019

Nora Tischler. 12 August 2020, 13 August 2020

Giulio Tononi. 12 January 2012, 8 January 2016, 21 January 2016, 29 February 2016, 10 March 2021, 20 October 2021, 21 October 2021, 22 October 2021

Vitaly Vanchurin. 31 August 2021

Marina Vegué Llorente. 28 June 2019, 9 November 2020, 10 November 2020

Jan Walleczek. 4 April 2018, 7 September 2019

David Wolpert. 2 April 2017, 12 June 2018

Lenka Zdeborová. 21 February 2019, 18 November 2019

Anton Zeilinger. 1 April 2011

ACKNOWLEDGMENTS

IN EXPLORING THIS TOPIC, I found myself entering a whole new community of scholars in physics, neuroscience, philosophy of mind, machine learning, and artificial intelligence. They are remarkable people—welcoming, generous with their time, and fun to hang out with.

A number of people reviewed passages of my manuscript: Scott Aaronson, Raphael Bousso, Časlav Brukner, Craig Callender, Dennis Dieks, Karl Friston, Christopher Fuchs, Ivette Fuentes, Andrew Haun, Erik Hoel, Donald Hoffman, Lorenzo Maccone, Kelvin McQueen, Ekaterina Moreva, Markus Müller, Renato Renner, Dan Roberts, Warrick Roseboom, Carlo Rovelli, Jascha Sohl-Dickstein, Aephraim Steinberg, Jun Tani, Giulio Tononi, David Wallace, and Wanja Wiese.

In addition to those I quote directly (listed in the bibliography), I'd also like to thank Olga Afanasjeva, Sean Carroll, Jitka Čejková, David Chalmers, Ian Durham, Ben Goertzel, Jakob Hohwy, Piet Hut, Pankaj Joshi, Johannes Kleiner, Hedda Mørch, Paavo Pylkkänen, Marek Rosa, and Howard Wiseman for amazing discussions.

I fell in with YHouse, an entertaining and erudite discussion group on neuroscience and AI in New York organized by Ayako

Fukui, Piet Hut, Sean Sakamoto, Caleb Scharf, and Olaf Witkowski. We were joined on occasion by Deepak Chopra and came to admire how he connects science to larger human concerns. During the pandemic lockdown, I kept up my intellectual connections through the Mathematical Consciousness Science seminar series, organized by Xerxes Arsiwalla, Johannes Kleiner, Robin Lorenz, Joanna Szczotka, and Sean Tull; the Harvard Philosophy of Science Club, organized by Jacob Barandes and Jonathan Haefner; and MIT's Program in Science, Technology, and Society, where I am an unpaid research affiliate.

I wouldn't have been able to attend conferences and visit scientists at their laboratories without the generous support of the Fetzer Franklin Fund Pioneers Award, for which I owe a debt of gratitude to Bruce Fetzer and Jan Walleczek. Jimyo Ferworn expertly handled the logistics.

The Foundational Questions Institute (FQxI) covered my travel costs to its international conferences in Banff in 2016 and Tuscany in 2019. I'm grateful to Anthony Aguirre and Max Tegmark for inviting me and to Kavita Rajanna for handling the logistics. FQxI, through a grant in its Consciousness in the Physical World program, also paid my way to the Mind and Agency in the Foundations of Quantum Physics conference at Chapman University in 2022. Many thanks to Kelvin McQueen for inviting me and to Tate Renville for handling logistics. I wrote two blog posts for FQxI's website, commissioned by Theiss Research through an FQxI minigrant.

Araya Brain Imaging covered my travel costs to the AI and Society Symposium in Tokyo in 2017. I'm grateful to Ayako Fukui and Ryota Kanai for inviting me.

The AGI Society covered my travel costs to attend the Human-Level AI conference in Prague in 2018. Olga Afanasjeva and Matthew Iklé made it all possible.

The Monash University Networks of Excellence grant covered my travel costs to attend the Causation, Complexity, and Consciousness workshop in Greece in 2018. I'm grateful to Tim Bayne and Nao Tsuchiya for inviting me and to Jasmine Walter for handling the logistics.

Robbert Dijkgraaf hosted me at the Institute for Advanced Study in Princeton as a Director's Visitor in 2018 and 2019. Many thanks to Natalie Wolchover and Nima Arkani-Hamed for suggesting my name and to Josephine Faass for making it possible.

The Institute for Pure and Applied Mathematics (IPAM) at UCLA covered my travel costs to the Using Physical Insights for Machine Learning workshop in 2019. I'm grateful to the organizing committee—Yann LeCun, Matthias Rupp, Lenka Zdeborová, and Riccardo Zecchina—for inviting me, to the program manager, Emily Roland, for handing the logistics, and to Kyle Cranmer for letting me know about the event. IPAM is supported by the National Science Foundation (Grant no. DMS-1925919).

The Wisconsin Institute for Sleep and Consciousness made it possible for me to visit in 2021. Many thanks to Giulio Tononi and Larissa Albantakis for hosting me and to Jonathan Lang for sorting out the logistics.

I'm grateful for media-relations officers for smoothing my way at various institutions, notably Anastasia Golovina at SingularityNET, Ali-Rae Hunt at D-Wave, Will Millership at GoodAI, and Tomomi Okubo at the Okinawa Institute of Science and Technology.

I started to explore the ideas in this book in articles for *Aeon*, *Nautilus*, *Psychology Today*, *Quanta*, *Science*, *Scientific American*, and *Spectrum News* magazines, as well as for NBC News MACH and the FQxI blog. Many thanks to my editors: Tim Appenzeller, Kevin Berger, Seth Fletcher, David Freeman, Eric Hand, Lybi Ma, Zeeya Merali, Michael Moyer, Kristin Ozelli, Liz Peterson, Corey Powell, Michael Segal, Gary Stix, Pamela Weintraub, and Ingrid Wickelgren.

I couldn't have run the marathon of writing a book if not for support from my fellow science writers, especially Mark Alpert, Anil Ananthaswamy, Steve Ashley, Dan Falk, Annaka Harris, Clara

Moskowitz, Philip Yam, and Lina Zeldovich. A special shout-out to Amanda Gefter for reading my whole manuscript and saving me from myself numerous times.

I wrote much of this book at local coffee shops, notably Legacy Coffee in Montclair, New Jersey, where Dewar and Sinéad MacLeod kept me well caffeinated and informed about local punk music, and 23 Skiddoo in Bloomfield, New Jersey, where Hodge Halili baked the best peanut-butter-cup cookies on Earth.

Where would I be without the team at Farrar, Straus and Giroux? Eric Chinski encouraged me to explore these themes long before I thought myself ready, and Ian Van Wye—well, let's just say that I owe him a lifetime of free beer for the pains he took to make the manuscript make sense. Christina Nichols was the copy editor that every writer dreams of, attending not just to errant punctuation but also to conceptual murkiness and narrative discontinuity. Thanks also to the proofreaders, Tanya Heinrich and Judy Kiviat, as well as to Patrice Sheridan, who designed the page layout, and Thomas Colligan, who came up with the awesome cover.

Lucy Reading-Ikkanda gets five stars for turning my rough sketches into finished illustrations. Phil Cantor, an old friend and master photographer, spent a day with me in his studio and at various locations shooting the jacket photo and other images. My agent, Susan Rabiner, ensured the project made sense and handled all the behind-the-scenes logistics.

My brother, Bret Musser, and brother-in-law, Jonathan Schaffer, have always been inspirations for me. Both are formidable scholars and entertaining dinner companions who can pivot from intellectually deep conversation to silly jokes. I dedicate this book to them.

My in-laws, Ann and Ben Schaffer, have always made me feel part of the family and sustained me with their love and pride.

I'm lucky to have inherited even a small fraction of my mother Judy's tennis skills and love of life. She has encouraged me in my writing ever since I first put crayon to construction paper.

Milo, the new little Schnauzer in our family, reminds me to take joy in the mundane. The sidewalk! The car! My human is home! Life is good!

The pandemic was especially hard on young people, yet my daughter, Eliana, discovered new interests and created online friendships that spanned a continent. At a time when many people tend to speak before they think, she is consistently thoughtful and deeply ethical. She makes me a better person in our every interaction. Especially when ice cream is involved.

Many physicists these days think the universe may ultimately be made of relations—that things may have no properties in isolation and are constituted by their interactions. My wife, Talia, explores in her academic work the same idea in the context of human culture. And she lives by this principle, too, shaping her every interaction to bring something new and special to everyone involved. How lucky I am to be one of the beneficiaries.

INDEX

Page numbers in *italics* refer to figures.

A Note About the Author

George Musser is an award-winning journalist, a contributing editor at *Scientific American*, a contributing writer at *Quanta*, and the author of *Spooky Action at a Distance*. He is the recipient of a Jonathan Eberhart Planetary Sciences Journalism Award from the American Astronomical Society and of the American Institute of Physics Science Communication Award for Science Writing. He was a Knight Science Journalism fellow at MIT and has appeared on *Today*, CNN, NPR, the BBC, and other outlets. He lives in Glen Ridge, New Jersey, with his wife, daughter, and Schnauzer. His website is georgemusser.com.